AF766343

Annotating new genes

Woodhead Publishing Series in Biomedicine

1. Practical leadership for biopharmaceutical executives
   J. Y. Chin
2. Outsourcing biopharma R&D to India
   P. R. Chowdhury
3. Matlab® in bioscience and biotechnology
   L. Burstein
4. Allergens and respiratory pollutants
   Edited by M. A. Williams
5. Concepts and techniques in genomics and proteomics
   N. Saraswathy and P. Ramalingam
6. An introduction to pharmaceutical sciences
   J. Roy
7. Patently innovative: How pharmaceutical firms use emerging patent law to extend monopolies on blockbuster drugs
   R. A. Bouchard
8. Therapeutic protein drug products: Practical approaches to formulation in the laboratory, manufacturing and the clinic
   Edited by B. K. Meyer
9. A biotech manager's handbook: A practical guide
   Edited by M. O'Neill and M. H. Hopkins
10. Clinical research in Asia: Opportunities and challenges
    U. Sahoo
11. Therapeutic antibody engineering: Current and future advances driving the strongest growth area in the pharma industry
    W. R. Strohl and L. M. Strohl
12. Commercialising the stem cell sciences
    O. Harvey

Published by Woodhead Publishing Limited, 2012

13

14 Human papillomavirus infections: From the laboratory to clinical practice
F. Cobo

15 Annotating new genes: From *in silico* to validations by experiments
S. Uchida

16 Open-source software in life science research: Practical solutions in the pharmaceutical industry and beyond
Edited by L. Harland and M. Forster

17 Nanoparticulate drug delivery: A perspective on the transition from laboratory to market
V. Patravale, P. Dandekar and R. Jain

18 Bacterial cellular metabolic systems: Metabolic regulation of a cell system with ^{13}C-metabolic flux analysis
K. Shimizu

19 Contract research and manufacturing services (CRAMS) in India: The business, legal, regulatory and tax environment
M. Antani and G. Gokhale

20 Bioinformatics for biomedical science and clinical applications
K-H. Liang

21 Deterministic versus stochastic modelling in biochemistry and systems biology
P. Lecca, I. Laurenzi and F. Jordan

22 Protein folding *in silico*: Protein folding versus protein structure prediction
I. Roterman

23 Computer-aided vaccine design
T. J. Chuan and S. Ranganathan

24 An introduction to biotechnology
W. T. Godbey

25 RNA interference: Therapeutic developments
T. Novobrantseva, P. Ge and G. Hinkle

26 Patent litigation in the pharmaceutical and biotechnology industries
G. Morgan

27 Clinical research in paediatric psychopharmacology: A practical guide
P. Auby

Annotating new genes

From *in silico* screening to experimental validation

Shizuka Uchida

WP
WOODHEAD
PUBLISHING

Oxford Cambridge Philadelphia New Delhi

Published by Woodhead Publishing Limited, 2012

Woodhead Publishing Limited, 80 High Street, Sawston, Cambridge, CB22 3HJ, UK
www.woodheadpublishing.com
www.woodheadpublishingonline.com

Woodhead Publishing, 1518 Walnut Street, Suite 1100, Philadelphia, PA 19102-3406, USA

Woodhead Publishing India Private Limited, G-2, Vardaan House, 7/28 Ansari Road, Daryaganj, New Delhi – 110002, India
www.woodheadpublishingindia.com

First published in 2012 by Woodhead Publishing Limited
ISBN: 978-1-907568-68-8 (print) and ISBN: 978-1-908818-12-6 (online)
Woodhead Publishing Series in Biomedicine ISSN 2050-0289 (print); ISSN 2050-2097 (online)

© S. Uchida, 2012

The right of S. Uchida to be identified as author(s) of this Work has been asserted by them in accordance with sections 77 and 78 of the Copyright, Designs and Patents Act 1988.

British Library Cataloguing-in-Publication Data: A catalogue record for this book is available from the British Library.

Library of Congress Control Number: 2012940176

All rights reserved. No part of this publication may be reproduced, stored in or introduced into a retrieval system, or transmitted, in any form, or by any means (electronic, mechanical, photocopying, recording or otherwise) without the prior written permission of the Publishers. This publication may not be lent, resold, hired out or otherwise disposed of by way of trade in any form of binding or cover other than that in which it is published without the prior consent of the Publishers. Any person who does any unauthorised act in relation to this publication may be liable to criminal prosecution and civil claims for damages.

Permissions may be sought from the Publishers at the above address.

The use in this publication of trade names, trademarks, service marks, and similar terms, even if they are not identified as such, is not to be taken as an expression of opinion as to whether or not they are subject to proprietary rights. The Publishers are not associated with any product or vendor mentioned in this publication.

The Publishers and author(s) have attempted to trace the copyright holders of all material reproduced in this publication and apologise to any copyright holders if permission to publish in this form has not been obtained. If any copyright material has not been acknowledged, please write and let us know so we may rectify in any future reprint. Any screenshots in this publication are the copyright of the website owner(s), unless indicated otherwise.

Limit of Liability/Disclaimer of Warranty
The Publishers and author(s) make no representations or warranties with respect to the accuracy or completeness of the contents of this publication and specifically disclaim all warranties, including without limitation warranties of fitness of a particular purpose. No warranty may be created or extended by sales of promotional materials. The advice and strategies contained herein may not be suitable for every situation. This publication is sold with the understanding that the Publishers are not rendering legal, accounting or other professional services. If professional assistance is required, the services of a competent professional person should be sought. No responsibility is assumed by the Publishers or author(s) for any loss of profit or any other commercial damages, injury and/or damage to persons or property as a matter of products liability, negligence or otherwise, or from any use or operation of any methods, products, instructions or ideas contained in the material herein. The fact that an organisation or website is referred to in this publication as a citation and/or potential source of further information does not mean that the Publishers nor the author(s) endorse the information the organisation or website may provide or recommendations it may make. Further, readers should be aware that internet websites listed in this work may have changed or disappeared between when this publication was written and when it is read. Because of rapid advances in medical sciences, in particular, independent verification of diagnoses and drug dosages should be made.

Typeset by Domex e-Data Pvt. Ltd., India

Contents

List of figures and tables

Figures

Tables

Acknowledgements

I would like to thank my family, my wife Petra, our daughter Erika and our son Leo for their support. I would like to thank Prof. Dr Dr Thomas Braun for his encouragement and support. My thanks go to my students, Ms Katharina Jenniches, Mr Pascal Gellert, Ms Piera De Gaspari, Ms Mizue Teranishi, Mr David John and Ms Yuliya Ponomareva. Without them, I would never thought about writing this book. I would also like to thank all of my collaborators. Finally, I would like to thank the following funding bodies for supporting me over the years: Yoshida Scholarship Foundation, Mochida Memorial Foundation for Medical and Pharmaceutical Research, Kerckhoff Foundation, Excellence Cluster Cardio-Pulmonary System (ECCPS), LOEWE Center for Cell and Gene Therapy (CGT) and University of Giessen-Marburg Lung Center (UGMLC).

About the author

Shizuka Uchida is a group leader at the Max Planck Institute for Heart and Lung Research (Bad Nauheim, Germany). His group consists of five PhD students and one masters student investigating the following three questions in biology: Are there tissue-enriched/specific (1) genes, (2) isoforms and (3) non-coding RNAs? To answer these questions, his group is employing computers using bioinformatics methods and performing biological experiments using model organisms ranging from fish (zebrafish) to transgenic mice. He is running a number of collaborations with medical doctors to conduct 'back to the bench' style studies to better treat patients.

After receiving a Bachelor of Science in biology and ancient Greek from Creighton University (Omaha, USA) in 1998, Uchida worked in business for four years. He returned to academia with Professor Akihiko Konagaya at the Japan Advanced Institute of Science and Technology (JAIST) (Ishikawa, Japan) in 2002. After a year and half under Associate Professor Kenji Satou, he started his PhD fellowship at the University of Louisville's Speed School of Engineering (Kentucky, USA) in 2003. To gain more hands-on wet lab experimentation, he returned to JAIST under academic advisor Prof. Toshifumi Tsukahara, an expert in biochemistry and molecular biology, focusing in particular on alternative splicing and microarray technology. After two years of PhD study, he obtained a number of grants and fellowships to study abroad. To further extend his interest in stem cell biology, he asked Prof. Dr Dr Thomas Braun at the Max-Planck-Institute for Heart and Lung Research (Bad Nauheim) to supervise his study. In 2007, Uchida defended his PhD degree at the JAIST. Since 2008, he has been supervising students to conduct research projects dealing mainly with developmental biology and adult stem cells in the fields of heart, lung and muscle development.

Shizuka Uchida was born and raised in Japan but spent his teenage years in Hawaii and Nebraska. Growing up in a melting pot of cultures (Hawaii) and educated in two extremes in educational systems [strict memorization (Japan) or complete individualism (USA)], he acquired a sense of balance. He believes in the idea of creating something new from old things, particularly fitting given the topic of this book.

He may be contacted at:

Max Planck Institute for Heart and Lung Research
Department of Cardiac Development and Remodelling
W.G. Kerckhoff-Institute
Ludwigstrasse 43
61231 Bad Nauheim
Germany
E-mail: Shizuka.Uchida@mpi-bn.mpg.de

1

Introduction

Everyone, at some point, must have been told: 'You look like your father (or mother).' A natural answer is: 'Of course, I am his/her son/daughter.' Such conversations have probably existed since humans were first able to communicate. However close the resemblance, though, we are not the same. We differ significantly from our parents in various aspects. And these differences are more evident if we compare ourselves to our friends, neighbors, teachers, professors, etc. So what creates these differences? Yes, genes.

At the turn of the 21st century, the draft of the human genome was completed. Regarded as a blueprint of a human being, the information contained in our DNA was considered to provide the ultimate answer to who we are. Fueled by the popular media, optimistic views have persisted that in the near future, we may be able to cure all diseases that threaten our lives. However, when these DNA sequences were analyzed, it was a surprise that the number of human genes is less than that of lower organisms (e.g. *Caenorhabditis elegans*). According to the GENCODE (http://www.gencodegenes.org/) Project, release 8, the number of human coding loci is 21,494. The key words here are 'coding loci'. Dr Jen Harrow, the Joint Head of Vertebrate Annotation at the Wellcome Trust Sanger Institute, avoids using the word 'gene' in this regard; there are protein-coding genes as well as non-coding genes present in the human genome.

A series of articles by 'Functional Annotation of Mammalian genome (FANTOM)' projects, the 'ENCyclopedia Of DNA Elements (ENCODE)' consortium and others have clearly indicated that a majority of the human genome is transcribed in the form of RNAs, yet only a few per cent of them fall under the category of protein-coding genes; the current estimate is that only ~1.2% of mammalian genomes encode for protein-coding genes (Clark *et al.*, 2011). Previously, non-coding RNAs (ncRNAs) were discarded as transcriptional noise and experimental error. However, with the discovery of micro-RNAs (miRNAs) and other types of ncRNAs [e.g. long non-coding RNAs (lncRNAs)], it is now clear that a minor proportion of RNAs are indeed translated as coding for proteins. The concept of ncRNAs is not new, for example as with ribosomal RNAs (rRNAs) and transfer RNAs (tRNAs), which are necessary for protein translation. The obvious question is thus is the definition of a 'gene' correct? In other words, how do we define a gene? This turns out to be very difficult to answer. Several years ago, *Nature* and *Science* published articles discussing this point (Pearson, 2006; Pennisi, 2007). An interesting article was also published by the members of the ENCODE consortium (Gerstein *et al.*, 2007), in which the authors provided an extensive review of the definitions of a gene used over the past century. After careful consideration, they proposed a new definition: 'A gene is a union of genomic sequences encoding a coherent set of potentially overlapping functional products.' As stated in their article, important aspects of this definition include that proteins and RNA products (e.g. ncRNAs) must be *functional*. With this definition, the authors proposed to move the long-standing biological question of 'what is a gene?' to 'what is a function?'. We therefore still have a long way to go to fully understand RNAs (both traditional genes and ncRNAs), which will require further biological experiments.

With the emergence of high-throughput techniques, such as microarrays and next-generation (deep) sequencing, the amount of biological data accumulated has expanded tremendously in the last few decades. These data now number not in the terabytes but in the petabytes (Stewart *et al.*, 2007). We clearly cannot interpret these data by hand, and numerous academic and commercial software products and databases are now available to facilitate their efficient analysis. However, the knowledge extracted from these data through such tools is still limited due to dependence on previous biological results and on connecting such results in an understandable manner. This dependency dismisses the presence of so-called 'functionally unknown genes'. These are putative genes whose presence is predicted from computational sequence analyses, but whose biological functions remain unknown; no published results are available for such genes (Uchida *et al.*, 2009).

For example, when a researcher performs microarray experiments to compare cells or tissues in one condition with another (e.g. operated and non-operated hearts), he or she hopes to identify known pathways in addition to new potential candidate genes that may interact with such pathways. However, in most cases, one may be able to identify several genes that are known to belong to the pathways and a number of genes with unknown function. When such unknown genes pop up in the results, an immediate response might be that these genes may possess functions related to a certain condition under study. However, it takes tremendous effort, money and time to elucidate the biological functions of such genes. To reduce the number of such genes to be studied further, other experiments, such as real-time reverse transcriptase PCR, must be performed to validate the results of the microarray data. Even then, the number of such functionally unknown genes is so large that understanding their biological functions would take several years or decades, if this were possible at all.

The presence of functionally unknown genes is evident from the data presented in Table 1.1, although the data do not include articles dealing with the large numbers of genes that are primarily due to the results of high-throughput technologies.

A similar number of functionally unknown genes was reported by Dr Jonathan Wren. He found that approximately 37% of the genes making up the human genome have no documented function (Wren, 2009). Furthermore, he indicated that the number of articles per gene was highly skewed to those already known. Studies focusing on the elucidation of novel genes appear to have decreased in recent years (Wren, 2009). Peña-Castillo and Hughes (2007) discussed this

Table 1.1 Uncharacterized genes (publications including more than 100 genes are excluded)

Organism	Number of Genes	Undescribed
Human	26,992	7,212 (26.7%)
Mouse	36,268	8,461 (23.3%)
Rat	26,136	12,884 (49.2%)
Chicken	18,691	13,906 (74.4%)
Zebrafish	33,359	18,760 (56.2%)

Data from Gellert et al. (2010).

point with regard to the best studied organism, yeast, explaining the large number of unknown genes as being due to genetic redundancy, a lack of strong phenotypes when genes are knocked out and a lack of evidence for bona fide genes. As the title of their article, 'Why are there still over 1000 uncharacterized yeast genes?', indicates, even in the most well studied organism, as of 2007, ~17% of genes remain unknown or uncharacterized (~6,000 in total).

With the further development of the 'omics' era in the last decade, worldwide efforts have focused on a new field of study termed 'Systems Biology' (Kitano, 2002). In this field, attempts are made to obtain, integrate and analyze complex biological data from various omics techniques (e.g. transcriptomics, proteomics and metabolomics) to understand biological phenomena as a collection of interconnected systems. However, no systems biological study is possible unless we know most, if not all, of the players in the systems under study. The presence of functionally unknown genes means that the current status of our biological knowledge is not sufficiently complete to conduct systems biology in an efficient and fruitful manner.

Recently, Dr Neil Greenspan published an opinion article titled 'Attributing functions to genes and gene products' (Greenspan, 2011), in which he stressed the importance of considering the function of a gene and its product in the sense of a mathematical function of multiple variables. By taking such a stand, it would, in my opinion, be nearly impossible to elucidate the function of an unknown gene with the currently available biological data. Thus, further biological experiments are essential.

The definition of genes remains of debate due to the presence of ncRNAs. Given that only ~1.2% of the transcribed genomes (i.e. RNAs) code for proteins, it is surprising that the biological functions of the majority of such protein-coding genes remain unknown. How is it possible to model a biological phenomenon without knowing all the players in the game? A car cannot be built with a black box of things whose functions are unknown! We as human beings are made of many things (far more than a car) that allow us to create things around us. How is it possible that we do not know ourselves (what we are made of) better than the things around us? This question is the motivation behind which I, together with my students and collaborators, started the *in silico* screening to identify genes that are evolutionarily conserved, tissue-enriched and whose functions are unknown (uncharacterized). This book was written to introduce you to these uncharacterized genes to better understand how we are made and what we are made of.

References

Clark MB, Amaral PP, Schlesinger FJ, *et al.* (2011) The reality of pervasive transcription. *PLoS Biol* 9(7): e1000625.

Gellert P, Jenniches K, Braun T, Uchida S. (2010) C-It: a knowledge database for tissue-enriched genes. *Bioinformatics* 26(18): 2328–33.

Gerstein MB, Bruce C, Rozowsky JS, *et al.* (2007) What is a gene, post-ENCODE? History and updated definition. *Genome Res* 17: 669–81. http://genome.cshlp.org/content/17/6/669.long.

Greenspan NS. (2011) Attributing functions to genes and gene products. *Trends Biochem Sci* 36(6): 293–7.

Kitano H. (2002) Systems biology: a brief overview. *Science* 295(5560): 1662–4.

Pearson H. (2006) Genetics: what is a gene? *Nature* 441(7092): 398–401.

Peña-Castillo L, Hughes TR. (2007) Why are there still over 1000 uncharacterized yeast genes? *Genetics* 176(1): 7–14.

Pennisi E. (2007) Genomics. DNA study forces rethink of what it means to be a gene. *Science* 316(5831): 1556–7.

Stewart CA, Müller M, Lingwall M. (2007) Progress towards petascale applications in biology: status in 2006. In: Lehner W, ed. *Euro-Par 2006 Workshops*. Berlin: Springer-Verlag, pp. 289–303.

Uchida S, Schneider A, Wiesnet M, *et al.* (2009) An integrated approach for the systematic identification and characterization of heart-enriched genes with unknown functions. *BMC Genomics* 10: 100.

Wren JD. (2009) A global meta-analysis of microarray expression data to predict unknown gene functions and estimate the literature–data divide. *Bioinformatics* 25: 1694–1701.

2

Databases and software to make your research life easier

Abstract: Through the accumulation of biological data, it is impossible to know everything about a gene or protein of interest without the aid of computers. To cope with the ever-increasing amount of data, various databases and software products have been developed to support the search for information. This chapter introduces a number of useful databases and software that you can use to save time and money for your ongoing and future projects.

Keywords: datasets, primary database, secondary database, software, web interface

2.1 Introduction

Data collection is a hallmark of biological experiments. Researchers conduct experiments to collect data in various formats: numbers, images, etc. These datasets form the foundation for testing one's hypothesis as well as planning for further experiments. To publish such results, these datasets must be analyzed properly to support or refute a hypothesis. With the emergence of high-throughput technologies (e.g. microarrays), the amount of biological datasets generated in one experiment is in the order of megabytes and approaching gigabytes for deep sequencing. As these datasets may be generated using public and government funding, it is often a requirement that they be disclosed to the public. To cope with this, various databases have been set up.

The best place to keep updated is probably the 'Database' issue of *Nucleic Acids Research* (http://nar.oxfordjournals.org/), which is published in January each year. According to the 18th issue of this journal, there were 1,330 databases and sources listed under the Molecular Biology Database Collection (http://www.oxfordjournals.org/nar/database/a) (Sayers *et al.*, 2011). In the following subsections, some of the representative databases and software that may be useful for the readers are explained in detail.

2.2 Primary databases

By definition, a 'primary database' is a type of database that consists of datasets derived experimentally (e.g. nucleotide sequences) and deposited by researchers. As the responsibility for data accuracy lies with the depositors (i.e. researchers), redundancies (overlaps) among the datasets as well as incompleteness of the data often make using such databases very time consuming.

2.2.1 Sequence databases

There are three main sequence databases: the DNA Data Bank of Japan (DDBJ) (http://www.ddbj.nig.ac.jp/index-e.html), EMBL Nucleotide Sequence Database (EMBL-Bank) (http://www.ebi.ac.uk/embl/) and GenBank (http://www.ncbi.nlm.nih.gov/genbank/). These function as repositories for nucleotide sequence data and form the foundations of the International Nucleotide Sequence Database Collaboration (INSDC) (http://www.insdc.org/). Although the names of the above databases indicate that they contain information about nucleotide sequences only, they also include amino acid sequences. They are all synchronized to exchange new and updated data on a daily basis to serve the research community.

As shown in Figure 2.1, most of these datasets consist of a very large amount of data. Therefore, it is not practical to download such datasets for private usage.

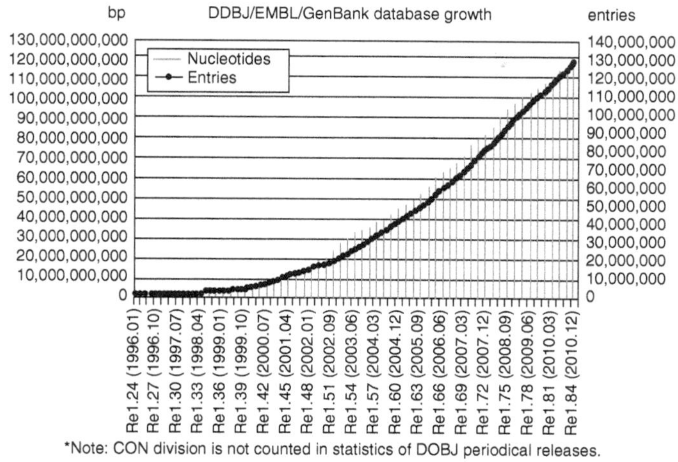

Figure 2.1 The amount of data deposited in INSDC (http://www.ddbj.nig.ac.jp/images/breakdown_stats/DBGrowth-e.gif). Accessed 25 February 2011

To find out more about how these databases are structured and connected, detailed information is available at http://www.insdc.org/documents/feature_table.html.

2.2.2 Structural databases

Originally, the Protein Data Bank (PDB) started as a repository for three-dimensional structural data of large biological molecules (e.g. proteins and nucleic acids). It is now maintained by an international organization called the Worldwide Protein Data Bank (wwPDB) (http://www.wwpdb.org/) consisting of the following members: Research Collaboratory for Structural Bioinformatics (RCSB) (http://www.rcsb.org/pdb/home/home.do), Protein Data Bank in Europe (PDBe) (http://www.ebi.ac.uk/pdbe/), Protein Data Bank Japan (PDBj) (http://www.pdbj.org/) and Biological Magnetic Resonance Data Bank (BMRB) (http://www.bmrb.wisc.edu/).

The types of datasets registered in the PDB are detailed in Table 2.1.

These data are especially useful for structural biologists who are looking for small molecules and drugs that might bind to the protein

Table 2.1 The type of datasets registered in the PDB. Accessed 25 February 2011 (http://www.pdb.org/pdb/statistics/holdings.do)

Experimental method	Proteins	Nucleic acids	Protein/NA complexes	Other	Total
X-ray	58,016	1,261	2,801	17	62,095
NMR	7,666	940	168	7	8,781
Electron microscopy	245	22	86	0	353
Hybrid	28	2	1	1	32
Other	132	4	5	13	154
Total	66,087	2,229	3,061	38	71,415

of interest to develop an antagonist or inhibitor for a particular biological phenomenon (e.g. disease). Furthermore, these data are necessary input information to predict how proteins fold themselves using simulations called 'molecular dynamics'.

2.2.3 Microarray databases

At present, DNA microarrays are the most commonly used mRNA profiling technique, and have proved to be an efficient tool for obtaining gene expression profiles at a particular time point. In the case of the Affymetrix Gene Arrays, only 50 ng of total mRNA is needed to perform an array experiment. In most cases, 2–3 days are required for the generation of the primary dataset. Through the standardization of protocols, systematic biases (noise) arising from microarrays (e.g. operator biases: Quackenbush, 2002; Uchida *et al*., 2005) have been minimized, which was supported by studies of the MicroArray Quality Control Consortium (MAQC Consortium, 2006).

Although some researchers prefer to disclose their microarray data (raw and/or analyzed) via their own homepage, many others deposit their datasets within public databases, such as ArrayExpress (http://www.ebi.ac.uk/arrayexpress/), Gene Expression Omnibus (GEO: www.ncbi.nlm.nih.gov/geo/) and Stanford Microarray Database (SDM: http://smd.stanford.edu/). Such public databases comply with the Minimum Information About a Microarray Experiment (MIAME: http://www.mged.org/Workgroups/MIAME/miame.html), which is a standard format for reporting microarray experiments to ensure that microarray data can be easily interpreted and their experiments to be reproduced to verify the results.

2.2.4 Proteomics databases

Considered to be the ultimate functional components of organisms' homeostasis and other important functions to sustain their life, proteins have attracted the attention of many researchers. With the rise of various techniques in the field of proteomics, it is now possible to study the signaling pathways at the level of proteins and their modifications (e.g. phosphorylation). There are various techniques and methods currently available in this field, such as Western blotting, two-dimensional gels and protein arrays (Rabilloud *et al.*, 2010; May *et al.*, 2011; Valledor and Jorrín, 2011). One of the most exciting and emerging technologies is mass spectrometry. The detailed procedures involved and the descriptions of different types of machines and approaches go well beyond the scope of this book. Readers are recommended to check the latest news and review articles from the PubMed database, which will be described in detail in the following subsections.

The development of high-end mass spectrometry has opened up a new horizon in the field of biology. Although the price of a mass spectrometer is beyond of the budget of a single lab, more and more research institutes and universities are installing such machines to cope with the increasing need to profile biological phenomena at the protein level. With the generation of increasing mass spectrometry data, several databases have been constructed to meet such needs (Table 2.2).

Table 2.2 List of proteomics databases

Name	URL	Description
Global Proteome Machine Organization	http://www.thegpm.org/	Mass spectral library and more information related to proteins and their modifications.
PeptideAtlas	http://www.peptideatlas.org/	Collection of peptides identified by tandem mass spectrometry proteomics experiments.
Peptidome	http://www.peptidome.jp/	Tandem mass spectrometry peptide and protein identification data.
PHOsphorylation SIte DAtabase (PHOSIDA)	http://www.phosida.de	Phosphorylation data generated in the lab of Dr Matthias Mann.
PRoteomics IDEntifications database (PRIDE)	http://www.ebi.ac.uk/pride/	Public repository for protein and peptide identifications maintained by the EMBL-EBI.
SIENA 2D-DATABASE	http://www.bio-mol.unisi.it/cgi-bin/2d/2d.cgi	2D-PAGE data.
TMIG-2DPAGE Proteome Database	http://www.proteome.jp/2D/J_index.html	2D-PAGE data focused on cellular aging and oxidative stress-induced cell damage.
WORLD-2DPAGE Constellation	http://world-2dpage.expasy.org/	2-D PAGE data.

2.2.5 Deep sequencing databases

Deep sequencing (also called 'second-generation sequencing' or 'next generation sequencing' to differentiate it from the Sanger sequencing method) has been a technology of choice in recent years. Several platforms are available, which I will not cover in this book (see excellent reviews by Mardis, 2008; Belncowe *et al.*, 2009; Wang *et al.*, 2009; Metzker, 2010). The sole purpose of deep sequencing is to read DNA sequences. However, when it is employed to mRNAs by constructing a cDNA library, the method itself is termed 'RNA-Seq'.

In principle, RNA-Seq can measure and quantify known genes as well as novel genes, isoforms and non-coding RNAs (ncRNAs). By employing this technology, it was found that over 90% of the human genes undergo alternative splicing to create at least one isoform from one gene (Wang *et al.*, 2008). Furthermore, it was this technology that allowed the presence of micro RNAs (miRNAs) to be established (Lu *et al.*, 2007; Friedlander *et al.*, 2008; Hafner *et al.*, 2008) and discovered new types of ncRNAs (Taft *et al.*, 2009) and long non-coding RNAs (lncRNAs) (Affymetrix ENCODE Transcriptome Project, 2009).

As in the case of nucleotide sequences, members of the INSDC maintain deep sequencing data: the DDBJ Sequence Read Archive (DRA) (http://trace.ddbj.nig.ac.jp/dra/index_e.shtml), NCBI Sequence Read Archive (SRA) (http://www.ncbi.nlm.nih.gov/sra) and EBI Sequence Read Archive (ERA) (http://www.ebi.ac.uk/ena/). However, due to the recent financial crisis, the SRA is closing; instead, other databases in the National Center for Biotechnology Information (NCBI) will take care of such deposits (e.g. RNA-seq and epigenomic data in the GEO database).

2.2.6 Databases with images

With advancements in the design of microscopes and the development of various imaging techniques, the field of cell biology has been rapidly expanding. Immunohistochemistry experiments and confocal or fluorescence microscopy are now standard techniques. The description of each of these goes well beyond the scope of this book; the information in Table 2.3 will direct readers to publicly available databases with images (some databases integrate information and images from other databases; therefore, they could also be considered as secondary databases, which are described in the next section).

Table 2.3 List of databases with images

Name	URL	Description
4DXpress	http://4dx.embl.de/4DXpress/guest/login.do;jsessionid=8EEFE98D009DD9A13F952BA2F135A880	Whole-mount *in situ* hybridization, antibody staining and transgenic animals for *Drosophila*, medaka, mouse and zebrafish.
Allen Brain Atlas	http://www.brain-map.org/	*In situ* hybridization and microarray of mouse and human brain.
American Society of Hematology (ASH) Image Bank	http://www.imagebank.hematology.org/	Images related to hematology.
Cell Centered Database (CCDB)	http://ccdb.UCSC.edu/index.shtm	High-resolution 2D, 3D and 4D data from light and electron microscopy.
Electron Microscopic Atlas	http://www.uni-mainz.de/FB/Medizin/Anatomie/workshop/EM/EMAtlas.html	Electron microscopic images of cells, tissues and organs.
Embryo Images Normal and Abnormal Mammalian Development	http://www.med.unc.edu/embryo_images/unit-welcome/welcome_htms/	Scanning electron micrographs of mouse and human embryos.
Embryonic Gene Expression Database for Biomedical Research Source (EMBRYS)	http://embrys.jp/embrys/html/MainMenu.html	Whole-mount *in situ* hybridization of murine embryos.
Eurexpress	http://www.eurexpress.org	Transcriptome atlas database for mouse embryo.
Gallus Expression *in situ* Hybridization Analysis (GEISHA)	http://geisha.arizona.edu/geisha/	*In situ* hybridization data for chicken embryos.
GenePaint.org	http://www.genepaint.org/Frameset.html	*In situ* hybridization on serial tissue sections of murine embryos.
Human Protein Atlas	http://www.proteinatlas.org//index.php	Immunohistochemical images of human tissues, cancers and cell lines.

Table 2.3 List of databases with images (*cont'd*)

Name	URL	Description
Multi-Dimensional Human Embryo	http://embryo.soad.umich.edu/index.html	Magnetic resonance images of human embryos.
Nordic immunohistochemical Quality Control (NordiQC)	http://www.nordiqc.org/	Immunohistochemistry images.
Stanford Tissue Microarray Database (TMAD)	http://tma.stanford.edu/cgi-bin/home.pl	Tissue microarray.
Web Atlas of Cellular Structures Using Light And Confocal Microscopy	http://www.itg.uiuc.edu/technology/atlas/	Subcellular structures and organelles using light and confocal microscopy.

2.2.7 Literature

One requirement of scientists is to disclose their findings to the public through publications. The major part of our grants derive from public taxes, and as such a minimum requirement should be to communicate through such means. There are many journals other than *Cell*, *Nature* and *Science* that are respected and important for one's particular research field, and the abstracts of most, but not all, such journals can be accessed freely through the Medical Literature Analysis and Retrieval System (MEDLINE). MEDLINE is a bibliographic database of life sciences and biomedical information operated by the US National Library of Medicine (NLM; http://www.nlm.nih.gov/pubs/factsheets/medline.htm). It contains over 18 million references to journal articles. It is linked to a freely available web search tool and database called PubMed (http://www.ncbi.nlm.nih.gov/pubmed/), which is integrated into the Entrez System (this system will be explained in the following subsections) and to the Medical Subject Headings (MeSH) (http://www.ncbi.nlm.nih.gov/mesh) to index articles using a vocabulary thesaurus, which allows for a more controlled and easy search for a particular topic.

PubMed provides a useful means to search for information. Through a function called My NCBI (http://www.ncbi.nlm.nih.gov/sites/myncbi/), searches and searched results can be saved. Furthermore, an automatic update through an email can be set up so that you can keep yourself up to date with the latest articles in your area of interest. For details of setting up My NCBI, please visit the following site: http://www.ncbi.nlm.nih.gov/sites/books/NBK3842/.

Given that most research has been conducted using money derived from public taxes, there has been fierce debate regarding charging fees to access published articles. Due to increased demands from the public, there are several so-called 'open access journals', which allow for online availability of published articles without financial, legal or technical barriers. Through the Directory of Open Access Journals (DOAJ; http://www.doaj.org/doaj?func=home&uiLanguage=en), open access journals can be searched. Connected to PubMed, there are databases that provide openly and freely accessible biomedical literature: PubMed Central (PMC; http://www.ncbi.nlm.nih.gov/pmc/) and UK PubMed Central (UKPMC; http://ukpmc.ac.uk/).

2.3 Secondary databases

A secondary database (also called 'metadatabase') consists of datasets that are derived from the analysis of primary data. In computer science, such analysis is often called 'data mining', a means by which to discover something new from existing datasets. Thus, a secondary database is also known as a 'knowledge database'. Often, they are structured such that the analyzed datasets can be downloaded in a text or table format. There are numerous secondary databases, so for the purposes of this book, I will cover only a limited number of the most relevant.

2.3.1 All-in-one databases

To facilitate understanding from biological datasets, it is of utmost importance to combine various information relating to a certain topic. Fueled by the development of internet technology, the exchange of primary datasets can be done very efficiently. As in any internet environment, one database is connected to another to form a web of connections. These connections can be intertwined to give multiple coverage of a gene or protein to better identify them. There are again many all-in-one databases, and I will focus in particular on their unique features (databases).

2.3.1.1 Entrez System

Established in 1988, the NCBI at the National Institutes of Health (NIH) maintains GenBank and provides data retrieval systems and computational resources and tools for the analysis of biological data (Sayers *et al.*, 2011). As a part of the United States National Library of Medicine (NLM), MEDLINE and PubMed are databases for articles published in the area of biosciences and medicine. To cope with such gigantic datasets, an integrated search and retrieval system to search the database contents of the NCBI website called 'Entrez' was created (Schuler *et al.*, 1996). As of April 2011, the Entrez system comprises over 40 molecular and literature databases. All the information contained in the system can be easily searched from its cross-database search page (http://www.ncbi.nlm.nih.gov/sites/gquery).

2.3.1.1.1 Gene

The Gene database (http://www.ncbi.nlm.nih.gov/gene) contains gene-centered information, such as nomenclature, chromosomal localization, gene products and their attributes, associated markers, phenotypes, interactions and links to various other information contained in the Entrez system and others (e.g. citations, sequences) (Maglott *et al.*, 2011). Each entry in the Gene database is assigned with a unique integer called GeneID (GI).

This GI is very helpful for programming purposes (which will be explained in detail in the next chapter). As most other secondary databases also use GIs as identifiers, various information obtained from many different databases can be linked at once using GI. Furthermore, the Entrez Gene database is linked to many other resources ('Links' on the right side of its main window), which makes it an excellent starting point to study a gene of interest.

The amount of data stored in the Gene database is enormous; over 7,300 taxa are covered. Up-to-date statistics for the Gene database can be accessed from the Gene Statistics site (http://www.ncbi.nlm.nih.gov/projects/Gene/gentrez_stats.cgi). The fact that all the contents of the database can be downloaded via its FTP site (ftp://ftp.ncbi.nlm.nih.gov/gene/) is an added advantage. The structures that are stored on the FTP site can be found in the README file (ftp://ftp.ncbi.nlm.nih.gov/gene/README). This information is important for programming purposes.

2.3.1.1.2 UniGene

UniGene (http://www.ncbi.nlm.nih.gov/unigene) (Wheeler *et al.*, 2003) is a database of non-redundant sets of transcript sequences. Each entry is classified under the same transcription locus (gene or expressed pseudogene). Conceptually, when researchers want to sequence an organism, they must first create a library [i.e. expressed sequence tag (EST) library] to be sequenced. This library will be constructed from total RNAs that are isolated from tissues or cells of the target organism. Once the sequencing is completed, the sequencing results will be mapped back to the genome to determine which gene is present or not. In other words, if a match is found, it is defined that this gene is expressed in the source tissue or cells. When such information is collected across various tissues and cells, one can determine in which body parts a gene is expressed or not. In addition, by combining various sequencing results, it is possible to count the number of sequences that match to the target gene. This number can be normalized by the total number of sequences per organ, to determine how often this gene appears (is expressed) in each organ. In the UniGene database, this normalized value is called 'transcripts per million (TPM)'.

As stated above, the information about gene expression is an interesting and useful feature of the UniGene database. Each UniGene ID ('Mm.290003' in Figure 2.2) is linked to such information

('EST Profile' and 'GEO Profiles'). When clicking [EST Profile], the page jumps to the one shown in Figure 2.2.

On the same page, the following statement can be found (not shown in Figure 2.2): 'EST profiles show approximate gene expression patterns as inferred from EST counts and the cDNA library sources (as reported by sequence submitters). Libraries known to be normalized, subtracted, or otherwise biased have been removed, but for a variety of reasons, EST counts may not be a true indication of gene activity.' In addition, because the UniGene database is updated between once a week and once a month (http://www.ncbi.nlm.nih.gov/UniGene/help.cgi?item=FAQ), the information about gene expression might change. Therefore, caution must be exercised when interpreting such results.

As one would expect, the expression of the *Myh6* gene is very specific to 'heart' [this gene is used as a marker for cardiomyocytes (heart muscle)]. The concept of 'specificity' is particularly relevant here, and will be discussed in detail in the following chapters. As shown in Figure 2.2, clicking on [more like this] displays the page shown in Figure 2.3.

As shown in Figure 2.3, there are 35 UniGene entries (or IDs; this word was chosen over 'genes' to meet the classification of the UniGene database, i.e. each UniGene entry represents a gene or pseudogene). Given that there are over 20,000 genes in the mouse genome, 35 UniGene entries seems a very small number. Therefore, the criteria used for this selection are very strict. For a UniGene entry to fall under this category, more than half of the GenBank sequences assigned to the cluster must come from the same source tissue. Interested readers can explore such a restricted expression using the following query term: '*(name of tissue/organ)'[restricted] AND (taxonomy ID)[taxid]*'. The latter part limits the search to a particular organism. Details about taxonomy IDs can be found in the Taxonomy database (http://www.ncbi.nlm.nih.gov/Taxonomy/). For example, the Taxonomy IDs for human, mouse and zebrafish are 9606, 10090 and 7955, respectively. Furthermore, a detailed description about how to search NCBI can be found at: http://www.ncbi.nlm.nih.gov/books/NBK3837/#EntrezHelp.Entrez_Searching_Options.

EST Profile - Mm.290003

NCBI » UniGene » EST Profile Viewer

Pubmed Nucleotide Protein Genome Structure Popset Taxonomy

Search UniGene

EST Profile

Mm.290003 - Myh6: Myosin, heavy polypeptide 6, cardiac muscle, alpha

Breakdown by Body Sites

		Mm.290003
adipose tissue	0	0 / 1564
adrenal gland	0	0 / 2044
bladder	0	0 / 13461
blood	0	0 / 16832
bone	0	0 / 32327
bone marrow	0	0 / 137491
brain	23	11 / 459991
connective tissue	0	0 / 19790
dorsal root ganglion	0	0 / 10186
embryonic tissue	18	12 / 664846
epididymis	0	0 / 2652
extraembryonic tissue	0	0 / 73200
eye	0	0 / 184900
fertilized ovum	0	0 / 26353
heart	3567	186 / 52131
inner ear	0	0 / 37947
intestine	0	0 / 84236
joint	0	0 / 17136
kidney	0	0 / 122453
liver	0	0 / 108957
lung	31	3 / 96442
lymph node	0	0 / 15170
mammary gland	0	0 / 302624
molar	0	0 / 3601
muscle	990	27 / 27256
nasopharynx	0	0 / 8067
olfactory mucosa	0	0 / 3281
ovary	0	0 / 53434
oviduct	0	0 / 3201
pancreas	0	0 / 105756
pineal gland	0	0 / 3914
pituitary gland	0	0 / 16965
prostate	0	0 / 29888
salivary gland	0	0 / 19345
skin	0	0 / 114950
spinal cord	0	0 / 21805
spleen	0	0 / 94430
stomach	0	0 / 29435
sympathetic ganglion	0	0 / 9045
testis	8	1 / 113759
thymus	0	0 / 116168
thyroid	0	0 / 8859

EST Profile - Mm.290003

tongue	0	0 / 9861
turbinate	0	0 / 1383
uterus	0	0 / 6842
vagina	0	0 / 5171
vesicular gland	0	0 / 2187

- Mm.290003 representation biased toward **heart** [more like this]

Breakdown by Developmental Stage

	Mm.290003	
oocyte	0	0 / 19423
unfertilized ovum	0	0 / 19710
zygote	0	0 / 27241
cleavage	0	0 / 27056
morula	0	0 / 37647
blastocyst	0	0 / 69305
egg cylinder	0	0 / 12083
gastrula	0	0 / 27624
organogenesis	63	8 / 125372
fetus	188	123 / 653835
neonate	338	35 / 103512
juvenile	51	15 / 291299
adult	35	35 / 995094

EST profiles show **approximate** gene expression patterns as inferred from EST counts and the cDNA library sources (as reported by sequence submitters). Libraries known to be normalized, subtracted, or otherwise biased have been removed, but for a variety of reasons, EST counts may not be a true indication of gene activity.

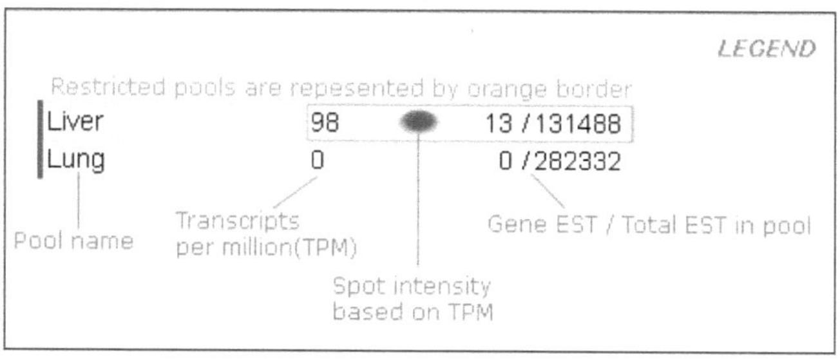

EST Profile - Mm.290003

Questions or Comments?
E-mail the NCBI Help Desk

FIRSTGOV

National Center for Biotechnology Information
U.S. National Library of Medicine
National Institutes of Health

U.S. DEPARTMENT OF HEALTH & HUMAN SERVICES

Disclaimer | Freedom of Information Act | Privacy Policy

http://www.ncbi.nlm.nih.gov/UniGene/ESTProfileViewer.cgi?uglist=Mm.290003[4/9/2011 6:49:57 AM]

Figure 2.2 Screenshot of the information about *Myh6* from the EST Profile database (http://www.ncbi.nlm.nih.gov/UniGene/ESTProfileViewer.cgi?uglist=Mm.290003). This has been modified to fit into one image. Accessed 1 May 2011

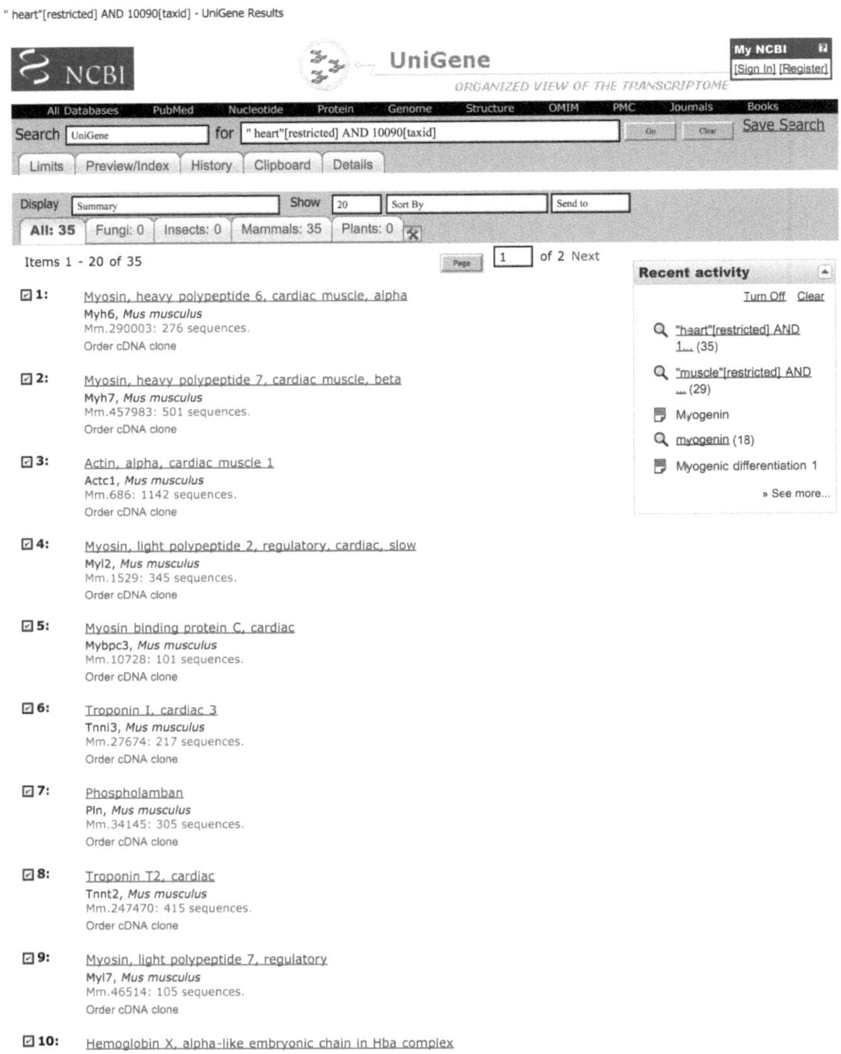
" heart"[restricted] AND 10090[taxid] - UniGene Results
NCBI
UniGene
ORGANIZED VIEW OF THE TRANSCRIPTOME
My NCBI
[Sign In] [Register]
All Databases PubMed Nucleotide Protein Genome Structure OMIM PMC Journals Books
Search UniGene for " heart"[restricted] AND 10090[taxid]
Go Clear
Save Search
Limits Preview/Index History Clipboard Details
Display Summary Show 20 Sort By Send to
All: 35 Fungi: 0 Insects: 0 Mammals: 35 Plants: 0
Items 1 - 20 of 35
Page 1 of 2 Next
Recent activity
Turn Off Clear
"heart"[restricted] AND 1... (35)
"muscle"[restricted] AND ... (29)
Myogenin
myogenin (18)
Myogenic differentiation 1
» See more...
1: Myosin, heavy polypeptide 6, cardiac muscle, alpha
Myh6, Mus musculus
Mm.290003: 276 sequences.
Order cDNA clone
2: Myosin, heavy polypeptide 7, cardiac muscle, beta
Myh7, Mus musculus
Mm.457983: 501 sequences.
Order cDNA clone
3: Actin, alpha, cardiac muscle 1
Actc1, Mus musculus
Mm.686: 1142 sequences.
Order cDNA clone
4: Myosin, light polypeptide 2, regulatory, cardiac, slow
Myl2, Mus musculus
Mm.1529: 345 sequences.
Order cDNA clone
5: Myosin binding protein C, cardiac
Mybpc3, Mus musculus
Mm.10728: 101 sequences.
Order cDNA clone
6: Troponin I, cardiac 3
Tnni3, Mus musculus
Mm.27674: 217 sequences.
Order cDNA clone
7: Phospholamban
Pln, Mus musculus
Mm.34145: 305 sequences.
Order cDNA clone
8: Troponin T2, cardiac
Tnnt2, Mus musculus
Mm.247470: 415 sequences.
Order cDNA clone
9: Myosin, light polypeptide 7, regulatory
Myl7, Mus musculus
Mm.46514: 105 sequences.
Order cDNA clone
10: Hemoglobin X, alpha-like embryonic chain in Hba complex

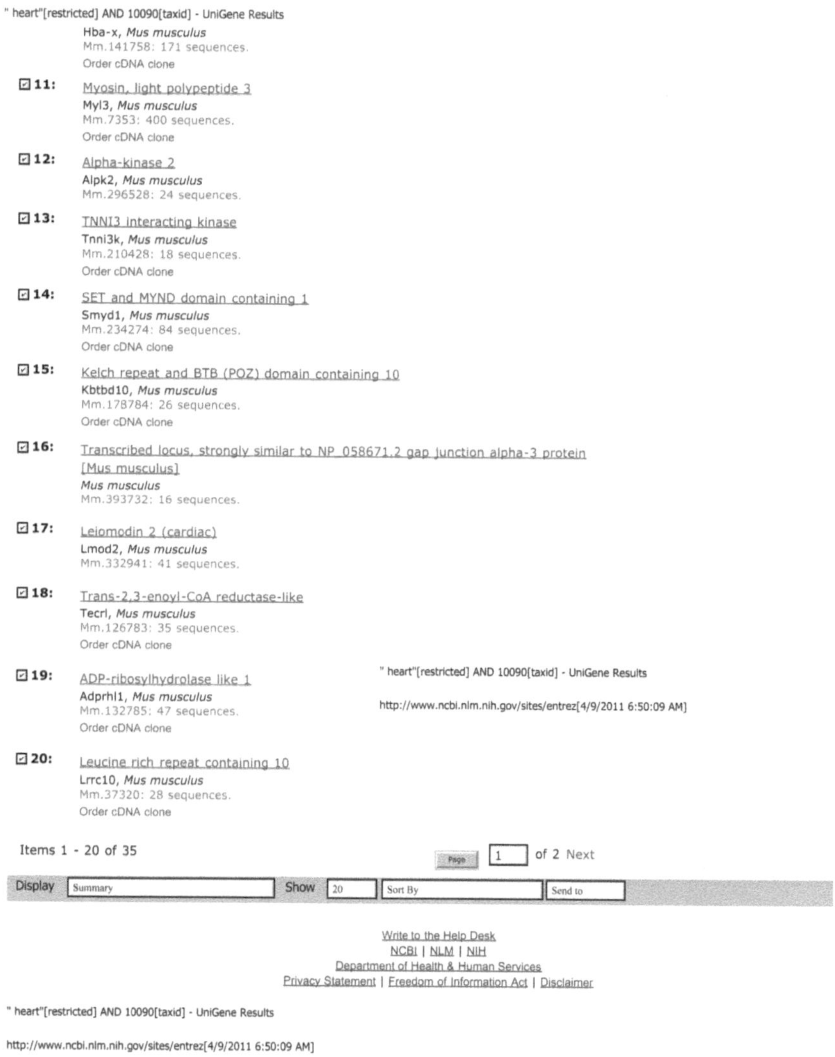

" heart"[restricted] AND 10090[taxid] - UniGene Results

Hba-x, *Mus musculus*
Mm.141758: 171 sequences.
Order cDNA clone

11: Myosin, light polypeptide 3
Myl3, *Mus musculus*
Mm.7353: 400 sequences.
Order cDNA clone

12: Alpha-kinase 2
Alpk2, *Mus musculus*
Mm.296528: 24 sequences.

13: TNNI3 interacting kinase
Tnni3k, *Mus musculus*
Mm.210428: 18 sequences.
Order cDNA clone

14: SET and MYND domain containing 1
Smyd1, *Mus musculus*
Mm.234274: 84 sequences.
Order cDNA clone

15: Kelch repeat and BTB (POZ) domain containing 10
Kbtbd10, *Mus musculus*
Mm.178784: 26 sequences.
Order cDNA clone

16: Transcribed locus, strongly similar to NP_058671.2 gap junction alpha-3 protein [Mus musculus]
Mus musculus
Mm.393732: 16 sequences.

17: Leiomodin 2 (cardiac)
Lmod2, *Mus musculus*
Mm.332941: 41 sequences.

18: Trans-2,3-enoyl-CoA reductase-like
Tecrl, *Mus musculus*
Mm.126783: 35 sequences.
Order cDNA clone

19: ADP-ribosylhydrolase like 1
Adprhl1, *Mus musculus*
Mm.132785: 47 sequences.
Order cDNA clone

" heart"[restricted] AND 10090[taxid] - UniGene Results

http://www.ncbi.nlm.nih.gov/sites/entrez[4/9/2011 6:50:09 AM]

20: Leucine rich repeat containing 10
Lrrc10, *Mus musculus*
Mm.37320: 28 sequences.
Order cDNA clone

Items 1 - 20 of 35 Page 1 of 2 Next

Display Summary Show 20 Sort By Send to

Write to the Help Desk
NCBI | NLM | NIH
Department of Health & Human Services
Privacy Statement | Freedom of Information Act | Disclaimer

" heart"[restricted] AND 10090[taxid] - UniGene Results

http://www.ncbi.nlm.nih.gov/sites/entrez[4/9/2011 6:50:09 AM]

Figure 2.3 Screenshot of the information about UniGene entries whose expressions are restricted to heart. Accessed 1 May 2011

2.3.1.1.3 SNP

Although humans have just 23 pairs of chromosomes, every human is unique (apart from identical twins). The differences are due to genetic variation: mutations occurring during development and gene copy number variations. When such variation occurs at the level of a single nucleotide among a population, it is called 'single-nucleotide polymorphism (SNP)'. As some individuals are more susceptible to certain diseases than others, SNPs are suspected as the cause and can be used as markers to screen for such diseases. There are several statistical methods to identify such SNPs [e.g. genome-wide association study (GWAS), linkage disequilibrium (LD)].

In the Entrez system, the dbSNP database (http://www.ncbi.nlm.nih.gov/projects/SNP/) contains SNPs, microsatellites, and small-scale insertions and deletions of genomes of various organisms (http://www.ncbi.nlm.nih.gov/projects/SNP/snp_summary.cgi).

2.3.1.1.4 OMIM

Initiated in the early 1960s by Dr Victor A. McKusick as a catalog of Mendelian traits and disorders in his book series, the Online Mendelian Inheritance in Man (OMIM; http://www.ncbi.nlm.nih.gov/omim) is a knowledge database to highlight the relationships between genotypes and phenotypes based on genes. It covers over 12,000 genes in relation to all known Mendelian disorders (http://www.ncbi.nlm.nih.gov/Omim/mimstats.html).

This database is very useful when looking for genes that are known to be related to diseases. As the ultimate goal of Biomedicine is to elucidate the mechanism of a particular disease in the hope of finding a cure, the information contained in the OMIM database provides a unique touch to the Entrez system.

2.3.1.1.5 Batch Entrez

One of the unique and useful features of the Entrez system is called the 'Batch Entrez' (http://www.ncbi.nlm.nih.gov/sites/batchentrez). Using GIs or accession numbers from the Nucleotide or Protein databases, a list of information contained in the Entrez system can be retrieved at once, and this can be saved as a file for further processing. To use the Batch Entrez, simply save a list of identifiers (e.g. GIs) in a file and upload it at the Batch Entrez website.

2.3.1.2 Ensembl

The Ensembl project (http://www.ensembl.org/) was launched in 1999 by the European Bioinformatics Institute (EBI) and the Wellcome Trust Sanger Institute to provide a centralized resource for genomic information relating to chordates. Information about other species (e.g. yeast and fly) is also provided to make comparative analyses. Unlike the Entrez system, the main feature of Ensembl is a genome browser. This genome browser includes many useful features. For a detailed description of these features, please refer to the tutorial materials provided by Ensembl: http://www.ensembl.org/info/website/tutorials/index.html.

Ensembl assigns its own unique ID called 'Ensembl ID'. This identifier starts with ENS (which stands for Ensembl; in the case of mouse, 'ENSMUS' is used) followed by 'G' for a gene, 'T' for a transcript, 'E' for an exon, 'P' for a peptide and 'F' for a family. These alphabets are followed by an 11-digit number, which does not change without a permanent change in its gene sequence or annotation. If you would like to know more about how an annotation is assigned, please refer to: http://www.ensembl.org/info/docs/genebuild/index.html.

A unique feature of Ensembl is the inclusion of accession numbers from various microarray probes and sequences to help facilitate discovery and provide linkages to the most popular high-throughput technology. Another unique feature is information about isoforms (Figure 2.4). As can be seen from this figure, there are six isoforms for *Myh6*. Being a genome browser, Ensembl provides a way to visually inspect all isoforms in one viewer (Figure 2.5).

To obtain a sequence (mRNA or protein), simply click on a hyperlinked Ensembl ID (shown in Figure 2.4), which jumps to the page with the sequence information. In the case of an mRNA

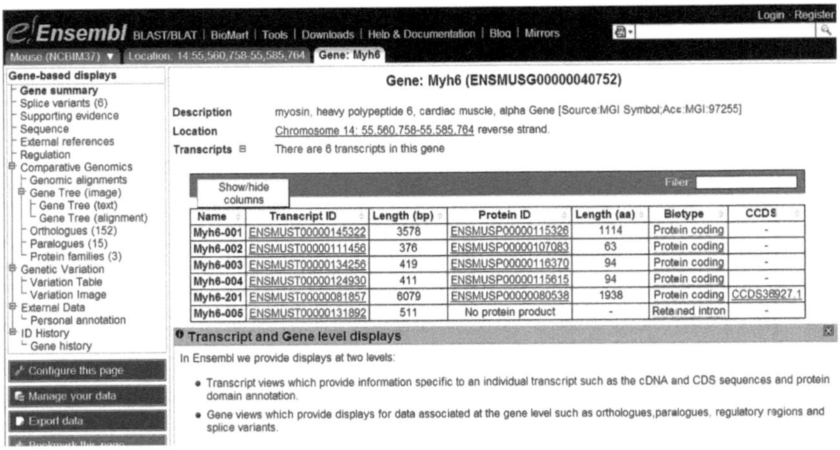

Figure 2.4 Screenshot of the information about *Myh6* in the Ensembl database (http://www.ensembl.org/Mus_musculus/Gene/Summary?g=ENSMUSG00000040752;r=14:55560758-55585764). Accessed 17 May 2011

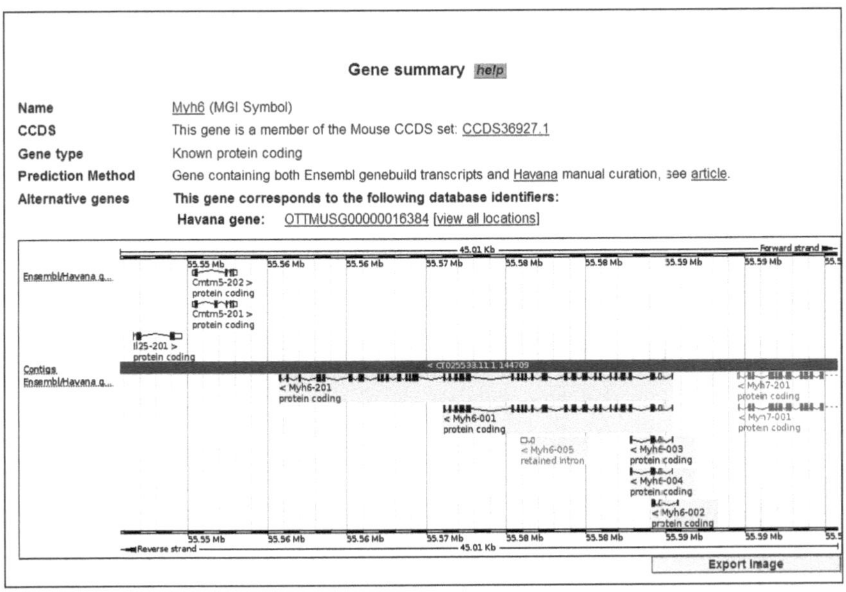

Figure 2.5 Screenshot of the genome browser for *Myh6* in the Ensembl database (http://www.ensembl.org/Mus_musculus/Gene/Summary?g=ENSMUSG00000040752;r=14:55560758-55585764). Accessed 17 May 2011

sequence, click on the 'Exons' shown at the left side of the window to display 5′ upstream, exons, introns and 3′ downstream sequences. This information is helpful when designing primers to perform RT-PCR experiments.

One of the two bodies behind Ensembl is the European Bioinformatics Institute (EBI). Therefore, it is no surprise that there is a function to retrieve gene-associated information without programming. This function is called 'BioMart' (http://www.biomart.org/). BioMart allows not only for the display of relevant information, but also the export of the searched information as text, html or Excel format to save the data and/or further process it for your specific needs.

2.3.1.3 UCSC Genome Browser

There are many genome browsers available worldwide, but the University of California, Santa Cruz (UCSC) Genome Browser (http://genome.ucsc.edu) is considered to be one of the most comprehensive. There are many features and functions that can be implemented in this genome browser. Below are some of the most useful features and functions:

- Customization of the browser to meet the needs of users. A sequence to be displayed can be of any size: a single DNA base up to the entire chromosome.
- Import of one's own data to compare with other data included in the UCSC Genome Browser using the Custom Tracks tool (http://genome.ucsc.edu/goldenPath/customTracks/custTracks.html).
- Support for various data formats, including deep sequencing data [e.g. BigBed, BigWig and Binary Alignment/Map (BAM)].
- A set of analysis tools to further process the data:
  - 'Table Browser' for data mining the information.
  - 'BLAT' (BLAST-like Alignment Tool) (Kent, 2002) for rapid searches by INDEXING the entire genome.

  - 'LiftOver' for translating genomic coordinates between assemblies.
  - 'Gene Sorter' for comparing genes based on expression profiles, protein homology, etc.
- Links to NCBI Map Viewer and the Ensembl Genome Browser for compatibility among viewers.

I strongly recommend trying this browser yourself to enjoy the pinnacle of Bioinformatics.

2.3.2 Secondary databases by topics

As there are literally hundreds of biological databases, it is not possible to cover all of them here. I have therefore selected databases based on my preferences (and apologize to the creators of the many excellent databases that are not mentioned) and briefly introduce them below.

2.3.2.1 Homologs

By definition, a homolog is a gene related to a gene in another species. More specifically, homologs can be classified as follows: (1) an ortholog is a gene evolved from a common ancestral gene by speciation – the functions of these two genes are similar, in most cases; (2) a paralog is a gene related to another gene by duplication within a genome – unlike orthologs, paralogs acquired new functions in the course of evolution. However, it is often difficult to determine if an evolutionarily conserved gene is an ortholog or paralog. Thus, throughout this book, I use 'homologs' to cover both 'orthologs' and 'paralogs'.

Most of the databases listing homologs are based on levels of sequence similarity at the gene and/or protein levels. One of the most comprehensive databases for homologs is 'Orthologous MAtrix (OMA)' (http://omabrowser.org). Initiated in 2004, the OMA database includes 1,000 genomes (827 bacteria, 103 eukaryotes and

70 archaea) (Altenhoff *et al.*, 2011). In the Entrez system, the 'HomoloGene' database (http://www.ncbi.nlm.nih.gov/homologene) provides information about homologs. Ensembl also has a database for homologs called 'Ensembl Compara' (http://www.ensembl.org/info/docs/api/compara/compara_schema.html). The strong sales point of this database is that it is programmable through the Application Program Interface (API). Information about Compara API can be found at: http://www.ensembl.org/info/docs/api/compara/index.html. Other useful homolog databases are: eggNOG (http://eggnog.embl.de/cgi_bin/show_input_page.pl), Hogenom (http://www.renabi.fr/article195.html), InParanoid (http://inparanoid.sbc.su.se/cgi-bin/index.cgi), MultiParanoid (http://multiparanoid.sbc.su.se/), OrthoDB (http://cegg.unige.ch/orthodb3), OrthoMCL (http://www.orthomcl.org/cgi-bin/OrthoMclWeb.cgi), PhylomeDB (http://phylomedb.org/) and Roundup (http://roundup.hms.harvard.edu/site/index.php).

A word of caution: this is not always a one-to-one relationship. Gene A in zebrafish might have several homologs in human. It is possible only to know exactly which one of several homologs is truly the one of interest based on biological experiments. We still have some way to go to comprehensively understand the evolutionary paths of organisms.

2.3.2.2 Organism-based secondary databases

Having described the databases for homologs, it is important to recognize that there are secondary databases available for model organisms to facilitate information exchange among researchers. Table 2.4 lists such databases for the major model organisms.

As I work extensively with mouse as a model organism, I will briefly introduce the Mouse Genome Informatics (MGI), which serves as a primary source of information. The MGI database includes hyperlinked information about genes, providing comprehensive coverage of their functions, genomic information, phenotypes and products (proteins). Most of the above information can be accessed through an easy-to-see summary page.

Table 2.4 List of organism-based databases

Name	URL	Description
GeneCards	http://www.genecards.org/	Human
H-InvDB	http://www.h-invitational.jp/index_jp.html	Human
HUGO Gene Nomenclature Committee (HGNC)	http://www.genenames.org/aboutHGNC.html	Human
Human Organized Whole genome Database (HOWDY)	http://howdy.biosciencedbc.jp/HOWDY/top_ja.pl	Human
Mouse Genome Informatics (MGI)	http://www.informatics.jax.org/	Mouse
Rat Genome Database (RGD)	http://rgd.mcw.edu/	Rat
BBSRC ChickEST Database	http://www.chick.manchester.ac.uk/	Chicken
Xenbase	http://www.xenbase.org/common/	Frog
Zebrafish Model Organism Database (ZFIN)	http://zfin.org/cgi-bin/webdriver?MIval=aa-ZDB_home.apg	Zebrafish
WormBase	http://www.wormbase.org/	*C. elegans*
FlyBase	http://flybase.org/	*Drosophila*
The Arabidopsis Information Resource (TAIR)	http://arabidopsis.org/	*Arabidopsis*
Saccharomyces Genome Database (SGD)	http://www.yeastgenome.org/	Budding yeast
Schizosaccharomyces pombe GeneDB (GeneDB)	http://old.genedb.org/genedb/pombe/	Fission yeast
EcoCyc	http://ecocyc.org/	*E. coli*

As there is a wide range of information provided in the MGI database, I will highlight those that are especially useful for the purpose of this book. Each MGI entry has its own unique ID called 'MGI ID'. The MGI ID for *Myh6* is 'MGI:97255'. This ID is linked to other publicly available databases (e.g. Entrez Gene, Ensembl, UniGene). These ID-to-ID linkages can be found on the FTP site: ftp://ftp.informatics.jax.org/pub/reports/index.html. Regarding such linkages, the information about transgenic mice is especially useful for researchers.

Phenotypic descriptions of *Myh6* transgenic mice are provided. Furthermore, if applicable, there is a hyperlink to images to allow visual inspections of such phenotypes. These images can be very helpful: 'A picture is worth a thousand words'. The image information is the main advantage of the MGI database, with images from published results, including the large-scale section *in situ* hybridization datasets from the Eurexpress atlas (http://www.eurexpress.org) (Diez-Roux *et al.*, 2011) and whole-mount *in situ* hybridization datasets from Gray *et al.* (2004).

There is a hyperlink to the International Mouse Knockout Consortium (http://www.knockoutmouse.org/), which aims to mutate all protein-coding genes in the mouse using a combination of gene trapping and gene targeting in C57BL/6 mouse embryonic stem cells. Through this link, a transgenic mouse line for a gene of interest may already be available for analysis. For specific usage, we recommend readers refer to a review article about the European Conditional Mouse Mutagenesis (EUCOMM) Program (Friedel *et al.*, 2007). Specific usage of such mouse lines for the purpose of this book will be described in detail in Chapters 5 and 6.

2.3.2.3 High-throughput phenotype screenings

The gold standard in the field of Developmental Biology is to find phenotypes that uncover the function of a gene of interest. Through the development of various techniques (e.g. gene targeting, knockdown by small hairpin RNAs), it is now possible to 'screen' for phenotypes in a high-throughput manner to elucidate the mechanism by which such a phenotype arises. Many secondary databases contain such phenotypic datasets (Table 2.5). Databases for mouse are not included in Table 2.5 as there is international effort to integrate various phenotypic databases called 'InterPhenome' (http://www.interphenome.org/).

As numerous databases for phenotypes are available, there are analysis tools and databases available to explore such datasets, including 'Mammalian Phenotype Enrichment Analysis (MamPhEA)'

Table 2.5 Secondary databases for phenotypes

Species	Database name	URL
Human	Database of Genotypes and Phenotypes (dbGaP)	http://www.ncbi.nlm.nih.gov/gap
Human and *Drosophila*	GenomeRNAi	http://rnai2.dkfz.de/GenomeRNAi/
Rat	PhenoMiner Database	http://rgd.mcw.edu/phenotypes/
C. elegans	WormMart	http://www.wormbase.org/biomart/martview
Rice	Oryza Tag Line (OTL)	http://urgi.versailles.inra.fr/OryzaTagLine
Rice	Rice Mutant Database (RMD)	http://rmd.ncpgr.cn/
Rice	Tos17	http://pc7080.abr.affrc.go.jp/phenotype
Rice	OryGenesDB	http://orygenesdb.cirad.fr/index.html
Barley	SCRI Barley Mutants	http://bioinf.scri.ac.uk/barley/
Maize	maizeGDB	http://www.maizegdb.org/rescuemu-phenotype.php
Tomato	Tomato Mutant Database	http://zamir.sgn.cornell.edu/mutants/
Tomato	LycoTILL	http://www.agrobios.it/tilling/index.html
Soybean	Soybean Mutation Database	http://www.soybeantilling.org/psearch.jsp
Influenza	Influenza Research Database	http://www.fludb.org/brc/home.do?decorator=influenza
Malaria	Rodent Malaria genetically modified Parasites (RMgmDB)	http://www.pberghei.eu/

(http://evol.nhri.org.tw/MamPhEA/) and 'PhenomicDB' (http://www.phenomicdb.de/).

One of the most exciting technologies that has emerged recently relates to 'zinc finger nucleases (ZFNs)', which allow targeted genomic manipulation in a variety of cell types from fruit flies to humans and also the creation of transgenic animals. Although ZFN technology is relatively new, the 'Zinc Finger Consortium' (http://www.zincfingers.org/default2.htm) has been established to ensure and promote its continued research and development. ZFN technology is promising to advance our understanding of Developmental Biology in the near future.

2.3.2.4 Categorizations of genes and proteins

As already mentioned, the major goal in the fields of Biology and Bioinformatics is to annotate genes by elucidating the underlining codes (e.g. laws of nature) of various organisms. To be able to annotate genes and proteins, it is of utmost importance to first categorize them. The following subsections introduce secondary databases with this aim in mind.

2.3.2.4.1 Gene ontology

In the field of microarray analysis, one of the most commonly used approaches to categorize differentially expressed genes [genes whose expression changes significantly from one condition (e.g. control group) to another (e.g. treated group)] is by Gene Ontology terms. The Gene Ontology (GO) project (http://www.geneontology.org/) provides a controlled vocabulary of terms (ontology) for describing gene products. GO covers three domains: (cellular component), the parts of a cell or its extracellular environment; (molecular function), the elemental activities of a gene product at the molecular level, such as binding or catalysis; and (biological process), operations or sets of molecular events with a defined beginning and end, pertinent to the functioning of integrated living units (cells, tissues, organs and organisms). The GO is structured as a directed acyclic graph. Each term has defined relationships to one or more of other terms in the same domain, and sometimes other domains. As the aim of GO is to provide standard vocabularies to researchers, the GO vocabularies are designed to be species-neutral to make them applicable to both prokaryotes and eukaryotes. Tools to explore the GO terms are provided by third parties listed at: http://www.geneontology.org/GO.tools.shtml.

2.3.2.4.2 UniProt

The UniProt Consortium (http://www.uniprot.org/) comprises the European Bioinformatics Institute (EBI), the Swiss Institute of Bioinformatics (SIB) and the Protein Information Resource (PIR) to provide a central repository of protein data. UniProt provides four core databases:

- UniProt Knowledgebase (UniProtKB): a protein database consisting of UniProtKB/Swiss-Prot (curated protein sequence database) and UniProtKB/TrEMBL (automatically annotated entries), which can be accessed through the Expert Protein Analysis System (ExPASy) server (http://expasy.org/sprot/).
- UniProt Archive (UniParc): a comprehensive and non-redundant database that contains most of the publicly available protein sequences in the world (http://www.ebi.ac.uk/uniparc/).
- UniProt Reference Clusters (UniRef): a collection of non-redundant clusters of protein sequences (http://www.ebi.ac.uk/uniref/).
- UniProt Metagenomic and Environmental Sequences (UniMES): a repository specifically developed for metagenomic and environmental data (http://www.uniprot.org/help/unimes).

Through the UniProt database, information about protein sequences and their functions can be freely obtained.

2.3.2.4.3 Databases for protein domains

A unique feature of proteins is that they can take three-dimensional structures, which include protein domains. A protein domain is any identifiable longer contiguous subsequence of a protein that can fold, function and exist independently of the rest of the protein chain or structure. As protein domains indicate the functional parts of the proteins in most cases, biochemical assays that biologists typically use are focused primarily on the protein domains of interest. Such experiments include the over-expression of a particular protein domain (gain-of-function) and deletion constructs targeting a particular protein domain (loss-of-function). Furthermore, some protein domains are conserved across species to provide similar functions among the homologous proteins. Therefore, there is significant interest in identifying conserved protein domains with functions that are important for the biological phenomenon of interest. There are various databases available to provide information about protein domains, the most well known being: Conserved Domain Database (CDD), http://www.ncbi.nlm.nih.gov/Structure/cdd/cdd.shtml; InterPro, http://www.ebi.ac.uk/interpro; Protein ANalysis THrough Evolutionary Relationships (PANTHER), http://www.pantherdb.org; Pfam, http://pfam.sanger.ac.uk/; and Simple Modular Architecture Research Tool (SMART), http://smart.embl-heidelberg.de.

2.3.2.4.4 Databases for molecular interactions

A protein often interacts with another protein to perform its function. At the systems level, each protein is connected to another to perform a complex task, such as respiration in the case of the citric acid cycle [also known as the tricarboxylic acid cycle (TCA cycle), the Krebs cycle or the Szent–Györgyi–Krebs cycle]. Such pathways are often represented at the gene level to make them easier to understand compared with proteins, which have different isoforms due to alternative splicing events. Table 2.6 provides a list of popular databases for molecular interactions.

There are many other useful molecular interaction databases open to the public, but which I did not cite in the above table. A comprehensive list of such databases can be found at: http://www.pathguide.org/.

Together with GO terms, information about molecular interactions, including signaling pathways, is something that one is interested in when analyzing high-throughput datasets, such as microarrays. One could explore the above databases to find relevant information, or upload a list of interesting genes or proteins into the interactive databases to obtain all the relevant information. The Batch Entrez described above provides such a function. A convenient and widely used system called the 'Database for Annotation, Visualization and

Table 2.6 List of molecular interaction databases

Name	URL	Description
Biocarta	http://www.biocarta.com/genes/index.asp	Pathways with nice graphics.
Biological General Repository for Interaction Datasets (BioGRID)	http://thebiogrid.org/	Curated biological database of protein–protein and genetic interactions.
Connections Maps	http://stke.sciencemag.org/cm/	Cellular signaling pathways and their relationships to one another.
Database of Interacting Proteins (DIP)	http://dip.doe-mbi.ucla.edu/dip/Main.cgi	Protein–protein interactions from experiments.
IntAct	www.ebi.ac.uk/intact/	Protein interaction data based on published results and direct user submissions.
Kyoto Encyclopedia of Genes and Genomes (KEGG)	www.genome.jp/kegg/	Knowledge base for systematic analysis of gene functions, linking genomic information with higher order functional information.
Pathway Interaction Database	http://pid.nci.nih.gov/	Curated collection of known biomolecular interactions.
Reactome	http://www.reactome.org/	Human pathways and reactions.
Search Tool for the Retrieval of Interacting Genes/Proteins (STRING)	http://string-db.org/	Experimental and predicted interaction information.
WikiPathways	http://www.wikipathways.org/index.php/WikiPathways	Wiki-style curated pathways.

Integrated Discovery (DAVID)' is also freely available through http://david.abcc.ncifcrf.gov/, and I strongly recommend that you make use of this database.

Not only can you search for molecular interactions, but GenMAPP (http://www.genmapp.org/) and Cytoscape (http://cytoscape.org/)

allow you to draw a pathway of your own. These freely available applications are designed to visualize gene expression and other genomic data on maps representing biological pathways and groupings of genes. The mapping can be done on well-known pathways as well as onto a newly created pathway. If you create an interesting pathway, you can make it available it through WikiPathways (http://www.wikipathways.org/index.php/WikiPathways).

2.3.2.5 ncRNAs

By definition, ncRNAs are RNA molecules that are not translated into proteins. Before the discovery of miRNAs, which are single-stranded 20–25 nt long RNAs (Lee *et al.*, 1993), the term ncRNAs was exclusively used to describe transfer RNAs (tRNAs) and ribosomal RNAs (rRNAs), which are abundant and functionally important RNAs. After over a decade of intensive research, we now know that thousands of different ncRNAs [e.g. small nucleolar RNAs (snoRNAs), Piwi-interacting RNAs (piRNAs) and long ncRNAs (more than 1,000 nt in length; Carcinci, 2010)] exist in mammals. Some, but not all, of these ncRNAs are strongly conserved during evolution (Ponting *et al.*, 2009). Although thousands of ncRNAs have been identified, only a handful have been studied in detail.

Results from the 'Functional Annotation Of Mammalian genome (FANTOM)' consortium unveiled 'a new continent in the RNA world' (Carninci *et al.*, 2005; Katayama *et al.*, 2005). This, together with the discovery of miRNAs, probably represents the greatest paradigm shift in the last decade in the field of biology. Although the functions of the majority of newly discovered ncRNAs remain unknown, some were found to play an important role in the regulation of cellular functions. Table 2.7 provides details of ncRNA databases.

Increasing evidence suggests that 'junk' DNA and experimental noise are long gone; most of the genome is transcribed to make RNA. The development of such databases will facilitate further discoveries relating to ncRNAs in the field of transcriptomics.

Table 2.7 List of ncRNA databases

Name	URL	Description
Allen Brain Atlas (ABA) ncRNA Expression Database	http://jsm-research.imb.uq.edu.au/ABAncrna/abancrnadb.pl	ncRNAs linked to the Allen Brain Atlas.
Arabidopsis thaliana Small RNA Project	http://asrp.cgrb.oregonstate.edu/	miRNAs in *Arabidopsis thaliana.*
Functional Annotation of the mouse 3 (FANTOM3)	http://fantom3.gsc.riken.jp/index.html	Cap-Analysis Gene Expression (CAGE) data.
Functional RNAdb (fRNAdb)	http://www.ncrna.org/frnadb/	Long, medium, small and microRNAs.
Long non-coding RNA database (lncrna db)	http://lncrnadb.com/	Long ncRNAs.
miRBase	http://www.mirbase.org/	Published miRNA sequences and annotation.
NCode Long Non-coding RNA Database	http://escience.invitrogen.com/ncRNA/?	Information relating to the long ncRNA and mRNA sequences profiled by the NCode Human and Mouse Non-Coding RNA Arrays.
ncRNAimprint	http://rnaqueen.sysu.edu.cn/ncRNAimprint/	Mammalian imprinted ncRNAs.
Noncoding RNA database	http://biobases.ibch.poznan.pl/ncRNA/	Sequences and annotations of ncRNAs from 99 species.
Noncoding RNA database (RNAdb)	http://research.imb.uq.edu.au/rnadb/	Sequences and annotations for 10,000 ncRNAs.
Plant snoRNA Database	http://bioinf.scri.sari.ac.uk/cgi-bin/plant_snorna/home	snoRNAs in plants.
Rfam	http://rfam.sanger.ac.uk/	ncRNA families and other structured RNA elements.
RNase P Database	http://jwbrown.mbio.ncsu.edu/RNaseP/home.html	RNase P sequences and related information.
Sno/scaRNAbase	http://bioinfo.fudan.edu.cn/snoRNAbase.nsf	snoRNAs and scaRNAs.

Table 2.7 List of ncRNA databases (*cont'd*)

Name	URL	Description
snoRNA Database	http://lowelab.ucsc.edu/snoRNAdb/	Methylation information about snoRNAs.
snoRNABase	http://www-snorna.biotoul.fr/	Human H/ACA and C/D box snoRNAs.
Yeast snoRNA Database	http://people.biochem.umass.edu/fournierlab/snornadb/main.php	*Saccharomyces cerevisiae* H/ACA and C/D box snoRNAs
NONCODE	http://www.noncode.org/NONCODERv3/	All kinds of non-coding RNAs (except tRNAs and rRNAs).

2.3.2.6 Literature-based secondary databases

Text mining is the analysis of text sources. When it is directed at the literature, it is often called 'literature mining'. As in data mining (in the strict sense, both text and literature mining should be classified as sub-fields of data mining), the field of literature mining has gained momentum in recent years, leading to the creation of secondary databases for scientific articles (Lu, 2011). For interested readers, several excellent review articles are available (Altman *et al.*, 2008; Cohen and Hunter, 2008; Rzhetsky *et al.*, 2009). As the scientific literature serves as a foundation for increased knowledge, numerous secondary databases have been developed using text and literature mining techniques. The Bio-NLP resources database (BioNLPdb) (http://zope.bioinfo.cnio.es/bionlp_tools) (Krallinger *et al.*, 2008) keeps track of such databases and tools.

In my opinion, one of the most useful literature-based secondary databases is called 'information on hyperlinked proteins (iHOP)' (http://www.ihop-net.org/UniPub/iHOP/). Based on the idea of providing a network of genes and proteins from the scientific literature, Drs Robert Hoffmann and Alfonso Valencia, two leaders in the field of text mining, built a database to hyperlink between sentences and abstracts in PubMed and to make them searchable by a gene name. The database is equipped with a function to tackle synonym ambiguity, 'iHOP confidence value' score being used to cope with a gene with various names in different fields. The user can save the searched results to build a network of their own based on genes; this function is called 'Gene Model', which allows for further knowledge discoveries from the user side. One of the unique features of the iHOP database is that the searched results can be superimposed onto published large-scale datasets (e.g. yeast-two hybrid data for protein–protein interactions) to facilitate the simultaneous analysis of novel and existing data.

Fueled by current trends within social media, a Wikipedia (http://www.wikipedia.org/)-like database was developed to provide context-based information to the public. This database is called 'WikiGenes' (http://www.wikigenes.org/?referer-ignore=1), which was developed and is maintained by Dr Robert Hoffmann. Conceptually, WikiGenes should represent the ultimate form of a knowledge database as it incorporates the ideas and evidence generated by various researchers. A word of caution: it is quite common to find that one hypothesis about a gene function proposed by a researcher might not be applied to the same gene under different biological phenomena due to multi-functionalities of the target gene. You might therefore find inconsistent (but not inaccurate) information for some genes.

2.4 Datasets to keep an eye out for

In the rest of this chapter, I would like to briefly describe datasets that I consider to be important for the purpose of this book; that is, to annotate new genes.

2.4.1 Genomics datasets

Through the emergence of deep sequencers, it is now possible to sequence an organism of interest far more quickly than ten years ago. One interesting attempt in this regard is the 'Genome 10K Project' (http://genome10k.org/), an international consortium to assemble a genomic zoo by sequencing a collection of DNA sequences for 10,000 vertebrate species obtained from zoos, museums and university collections around the world. If this is successful, then we could expect to see an avalanche of sequencing data, which will dynamically move the field of Evolutionary Biology towards understanding the evolutionary paths of many species.

With the goal to build a comprehensive parts list of functional elements in the human genome, the Encyclopedia Of DNA Elements (ENCODE) project (http://www.genome.gov/10005107#4) employs a variety of assays and methods (e.g. RNA-seq, ChIP-seq) to identify functional elements and to annotate such regions to understand genes better. The mouse version of ENCODE is in development (http://genome.ucsc.edu/encode/dataSummaryMouse.html). Furthermore, initiated in 2007, the Model Organism ENCyclopedia Of DNA Elements (modENCODE) project released its first set of results for fly and worm genomes in 2010 (http://www.modencode.org/). The data from ENCODE and modENCODE can be accessed through GEO (http://www.ncbi.nlm.nih.gov/geo/info/ENCODE.htm) and can be explored through the UCSC Genome Browser (http://genome.ucsc.edu/ENCODE/).

By definition, epigenetics is the study of inherited changes of gene expression without alterations of primary DNA sequences (Bird, 2007). Increasing evidence in recent years clearly indicates that the epigenetic status of a genome is important for various biological phenomena. Therefore, it is no surprise that various so-called 'epigenome projects' have been launched recently. Some projects [e.g. the Human Epigenome Project (HEP) (http://www.epigenome.org/)] are funded with hefty grants to characterize epigenetic status in various conditions. Such projects will undoubtedly add more data (the order of terabytes) to the already 'enough' biological data in the world.

2.4.2 Transcriptomics datasets

As described above, microarray technology has dominated the area of transcriptomics research for the last two decades. Those datasets dealing with tissues and/or organs are of particular interest for us, for example:

- BioGPS (http://biogps.gnf.org/): various tissues and organs of human and mouse.
- StemBase (http://www.stembase.ca): mostly dealing with stem cells.
- Genotype-Tissue Expression (GTEx) (https://commonfund.nih.gov/GTEx/): the Common Fund's program aims at studying human gene expression and regulation in multiple tissues, providing valuable insight into the mechanisms of gene regulation and, in the future, its disease-related perturbations.

Serial analysis of gene expression (SAGE) was developed in the early 1990s to detect mRNAs in a sample of interest by utilizing small tags that correspond to fragments of those transcripts (Velculescu *et al.*, 1995). Unlike microarrays, SAGE does not require pre-existing sequences for probe construction, which allows for *de novo* identification of novel genes. After sequencing of all tags, the sequences are mapped back to the mRNA sequence database (e.g. SAGEmap; http://www.ncbi.nlm.nih.gov/projects/SAGE/) to identify the corresponding genes as well as to count how often each tag appears, to give a quantification of the expression levels in comparison with a control sample (Wang, 2007). This identification of sequenced tags poses a problem due to their short sequences. Therefore, since its introduction in 1995, the length of small fragments of sequences has increased from 10 (short SAGE) to 17 (long SAGE), 21 (Robust-LongSAGE tag; Gowda *et al.*, 2004) and 26 (Super-SAGE tag; Matsumura *et al.*, 2005). Of the SAGE data, one study is particularly interesting, namely the 'Mouse Atlas of Gene Expression' (http://www.mouseatlas.org/) (Siddiqui *et al.*, 2005), which we used for our 'C-It' database (http://C-It.mpi-bn.mpg.de) (Gellert *et al.*, 2010). With the goal to provide a quantitative and comprehensive atlas of gene expression during the development of mice, the Mouse Atlas of Gene Expression project used SAGE technology to catalog the gene expression profiles of C57BL/6J mice during embryonic development as well as at adulthood. These datasets are particularly important to understand gene expression patterns during the development of murine embryos.

Recent advances in deep sequencing technology will clearly generate more and more data in the near future. The speed of progress is far surpassing what we imagined ten years ago. It is now possible to sequence the whole genome of a lower organism in a matter of days. Interesting approaches using deep sequencing, such as metagenomics, have revolutionized the way research is carried out. Due to such momentum in the generation of enormous amounts of data, the main current challenge is to develop a fast and accurate method to analyze such data, a daunting task for bioinformaticians.

2.4.3 Proteomics datasets

The waves of high-throughput and automated approaches have revolutionized the way we do research. In the field of Proteomics, such approaches are especially useful to find protein–protein interactions (PPIs), for example the 'yeast-two-hybrid' approach. A yeast-two-hybrid method is based on the following concept: using two hybrid molecules ('bait' protein, which is fused to a transcription factor; and 'prey' protein, which is fused to a related transcription factor), if the bait and prey proteins interact, then the two transcription factors fused to these two proteins are also brought into close proximity to each other. This results in the production of a specific signal (e.g. LacZ), which indicates an interaction has taken place. Ultimately, we are able to conclude that two proteins (bait and prey) interact with each other. Although there are several caveats to bear in mind with this method (e.g. false positive/negative interactions), the method itself has been standardized so that by using robots, it is possible to automate most of it in a high-throughput manner. Some organizations and institutes disclose yeast-two-hybrid datasets through their websites in the form of a searchable database [e.g. the 'Drosophila Interactions Database (DroID)' (http://www.droidb.org/) for PPIs in fruit fly and the 'Human Unidentified Gene-Encoded (HUGE)' protein database (http://www.kazusa.or.jp/huge/) for PPIs of human KIAA proteins], whose datasets can be downloaded through their websites.

2.5 Conclusion

Overwhelmed? I bet you are!! The list of databases and datasets surveyed here is nowhere near complete, and I apologize to those authors and researchers who have generated databases not cited here. There are simply too many out there to be able to cover them all. How much money (in the form of grants, either governmental or private) has been spent to generate such data and why are they not being used more often? This is exactly the motivation behind *in silico* screening. The following chapters discuss our approach to *in silico* screening in detail.

References

Affymetrix ENCODE Transcriptome Project. (2009) Affymetrix ENCODE Transcriptome Project; Cold Spring Harbor Laboratory ENCODE Transcriptome Project (2009) Post-transcriptional processing generates a diversity of 5'-modified long and short RNAs. *Nature* 457(7232): 1028–32.

Altenhoff AM, Schneider A, Gonnet GH, Dessimoz C. (2011) OMA 2011: orthology inference among 1000 complete genomes. *Nucleic Acids Res* 39(Database issue): D289–94.

Altman RB, Bergman CM, Blake J, *et al.* (2008) Text mining for biology – the way forward: opinions from leading scientists. *Genome Biol* 9(Suppl. 2): S7.

Blencowe BJ, Ahmad S, Lee LJ. (2009) Current-generation high-throughput sequencing: deepening insights into mammalian transcriptomes. *Genes Dev* 23(12): 1379–86.

Bird A. (2007) Perceptions of epigenetics. *Nature* 447(7143): 396–8.

Carninci P. (2010) RNA dust: where are the genes? *DNA Res* 17(2): 51–9.

Carninci P, Kasukawa T, Katayama S, *et al.*; FANTOM Consortium; RIKEN Genome Exploration Research Group and Genome Science Group (Genome Network Project Core Group) (2005) The transcriptional landscape of the mammalian genome. *Science* 309(5740): 1559–63.

Cohen KB, Hunter L. (2008) Getting started in text mining. *PLoS Comput Biol* 4(1): e20.

Diez-Roux G, Banfi S, Sultan M, *et al.* (2011) A high-resolution anatomical atlas of the transcriptome in the mouse embryo. *PLoS Biol* 9(1): e1000582.

Friedel RH, Seisenberger C, Kaloff C, Wurst W (2007) EUCOMM – the European conditional mouse mutagenesis program. *Brief Funct Genomic Proteomic* 6(3): 180–5.

Friedländer MR, Chen W, Adamidi C, *et al.* (2008) Discovering microRNAs from deep sequencing data using miRDeep. *Nat Biotechnol* 26(4): 407–15.

Gellert P, Jenniches K, Braun T, Uchida S (2010) C-It: a knowledge database for tissue-enriched genes. *Bioinformatics* 26(18): 2328–33.

Gowda M, Jantasuriyarat C, Dean RA, Wang GL (2004) Robust-LongSAGE (RL-SAGE): a substantially improved LongSAGE method for gene discovery and transcriptome analysis. *Plant Physiol* 134(3): 890–7.

Gray PA, Fu H, Luo P, *et al.* (2004) Mouse brain organization revealed through direct genome-scale TF expression analysis. *Science* 306(5705): 2255–7.

Hafner M, Landgraf P, Ludwig J, *et al.* (2008) Identification of microRNAs and other small regulatory RNAs using cDNA library sequencing. *Methods* 44(1): 3–12.

Katayama S, Tomaru Y, Kasukawa T, *et al.*; RIKEN Genome Exploration Research Group; Genome Science Group (Genome Network Project Core Group); FANTOM Consortium (2005) Antisense transcription in the mammalian transcriptome. *Science* 309(5740): 1564–6.

Kent WJ. (2002) BLAT – the BLAST-like alignment tool. *Genome Res* 12(4): 656–64.

Krallinger M, Valencia A, Hirschman L. (2008) Linking genes to literature: text mining, information extraction, and retrieval applications for biology. *Genome Biol* 9(Suppl. 2): S8.

Lee RC, Feinbaum RL, Ambros V. (1993) The *C. elegans* heterochronic gene lin-4 encodes small RNAs with antisense complementarity to lin-14. *Cell* 75(5): 843–54.

Lu C, Meyers BC, Green PJ. (2007) Construction of small RNA cDNA libraries for deep sequencing. *Methods* 43(2): 110–17.

Lu Z. (2011) PubMed and beyond: a survey of web tools for searching biomedical literature. *Database* (Oxford) 2011: baq036.

Maglott D, Ostell J, Pruitt KD, Tatusova T. (2011) Entrez Gene: gene-centered information at NCBI. *Nucleic Acids Res* 39(Database issue): D52–7.

MAQC Consortium, Shi L, Reid LH, Jones WD, *et al.* (2006) The MicroArray Quality Control (MAQC) project shows inter- and intraplatform reproducibility of gene expression measurements. *Nat Biotechnol* 24(9): 1151–61. http://www.ncbi.nlm.nih.gov/pubmed/16964229.

Hafner M, Landgraf P, Ludwig J, *et al.* (2008) Identification of microRNAs and other small regulatory RNAs using cDNA library sequencing. *Methods* 44(1): 3–12. http://www.ncbi.nlm.nih.gov/pubmed/18158127

Mardis ER. (2008) The impact of next-generation sequencing technology on genetics. *Trends Genet* 24(3): 133–41.

Matsumura H, Ito A, Saitoh H, *et al.* (2005) SuperSAGE. *Cell Microbiol* 7(1): 11–18.

May C, Brosseron F, Chartowski P, Schumbrutzki C, Schoenebeck B, Marcus K. (2011) Instruments and methods in proteomics. *Methods Mol Biol* 696: 3–26.

Metzker ML. (2010) Sequencing technologies – the next generation. *Nat Rev Genet* 11(1): 31–46.

Ponting CP, Oliver PL, Reik W. (2009) Evolution and functions of long noncoding RNAs. *Cell* 136(4): 629–41.

Quackenbush J. (2002) Microarray data normalization and transformation. *Nat Genet* 32(Suppl.): 496–501.

Rabilloud T, Chevallet M, Luche S, Lelong C. (2010) Two-dimensional gel electrophoresis in proteomics: past, present and future. *J Proteomics* 73(11): 2064–77.

Rzhetsky A, Seringhaus M, Gerstein MB. (2009) Getting started in text mining: part two. *PLoS Comput Biol* 5(7): e1000411.

Sayers EW, Barrett T, Benson DA, *et al.* (2011) Database resources of the National Center for Biotechnology Information. *Nucleic Acids Res* 39(Database issue): D38–51.

Schuler GD, Epstein JA, Ohkawa H, Kans JA. (1996) Entrez: molecular biology database and retrieval system. *Methods Enzymol* 266: 141–62.

Siddiqui AS, Khattra J, Delaney AD, *et al.* (2005) A mouse atlas of gene expression: large-scale digital gene-expression profiles from precisely defined developing C57BL/6J mouse tissues and cells. *Proc Natl Acad Sci USA* 102(51): 18485–90.

Taft RJ, Glazov EA, Cloonan N, *et al.* (2009) Tiny RNAs associated with transcription start sites in animals. *Nat Genet* 41(5): 572–8.

Uchida S, Nishida Y, Satou K, Muta S, Tashiro K, Kuhara S. (2005) Detection and normalization of biases present in spotted cDNA microarray data: a composite method addressing dye, intensity-dependent, spatially-dependent, and print-order biases. *DNA Res* 12(1): 1–7.

Valledor L, Jorrín J. (2011) Back to the basics: maximizing the information obtained by quantitative two dimensional gel electrophoresis analyses by an appropriate experimental design and statistical analyses. *J Proteomics* 74(1): 1–18.

Velculescu VE, Zhang L, Vogelstein B, Kinzler KW. (1995) Serial analysis of gene expression. *Science* 270(5235): 484–7.

Wang ET, Sandberg R, Luo S, *et al.* (2008) Alternative isoform regulation in human tissue transcriptomes. *Nature* 456(7221): 470–6.

Wang SM. (2007) Understanding SAGE data. *Trends Genet* 23(1): 42–50.

Wang Z, Gerstein M, Snyder M. (2009) RNA-Seq: a revolutionary tool for transcriptomics. *Nat Rev Genet* 10(1): 57–63.

Wheeler DL, Church DM, Federhen S, *et al.* (2003) Database resources of the National Center for Biotechnology. *Nucleic Acids Res* 31(1): 28–33.

3

How to program like a bioinformatician?

Abstract: Nowadays, it is impossible to analyze biological data without the help of computers. This is clearly the case for high-throughput datasets (e.g. microarrays, deep sequencing). Many students and researchers specializing in Biology use computers daily to analyze their results, but a lack of exposure to Computer Science means that usage is rather limited and depends very much on commercially and publicly available databases and software products. This chapter aims to introduce readers to the art of programming through Perl to make the daily tasks of researchers a bit easier.

Keywords: Cygwin, Perl, programming

3.1 Introduction

The majority of datasets that we deal with in Biology comprise text and numbers. With regard to biological datasets, significance is often determined at the levels of genes (both DNA and RNA) and proteins. 'Annotations' are therefore required. By definition, an 'annotation' is a description of what a gene is; in other words, what a DNA sequence represents – such as a gene. Unfortunately, this concept of annotations has caused problems among biologists. Through the emergence of various high-throughput techniques (e.g. microarrays, deep sequencing, mass spectrometry), the amount of data that can be generated in one experiment is already at the level of gigabytes. Due to this enormous increase in the number of datasets, it is now simply impossible to analyze them by hand. It is not possible to inspect such data values one-by-one by a single person without the aid of computational methods. This is exactly the point made in an issue of *Nature* [4 September 2008; 455 (7209)] with cover figure 'BIG DATA: Science in the petabyte era'. A more pressing problem is that the central dogma of Molecular Biology (i.e. DNA is transcribed to RNA, which is translated to proteins) has been modified due to the discovery of non-coding RNAs [hallmarked by microRNAs (miRNAs)]. To further complicate the matter, a recent deep sequencing study indicates that nearly every gene has at least one alternative isoform (Wang *et al.*, 2008), which can be expressed in a tissue-, development- or sex-specific manner and fulfill different or even opposing functions (Stetefeld and Ruegg, 2005; Wang *et al.*, 2008; Blekhman *et al.*, 2010). The point of interest here is that in order to interpret the biological data, it is necessary to take all traditional and current understanding together with previously published datasets to (data) mine the obtained datasets and to extract 'knowledge' (i.e. something that is meaningful to validate one's hypothesis and to perform further studies). This is impossible to do without the help of computers and computer programs.

Being a bioinformatician, I am often asked by students how difficult it is to program. Also, many biologists bring me their experimental results (mostly microarrays) and expect major findings to be extracted. If they want their biological experimental datasets to be analyzed by bioinformaticians, it is absolutely essential that such experts are involved from the planning stage of the experiments to obtain maximum output from limited resources. Alternatively, I tell them that it is not difficult to program. In this chapter, I will guide the readers through the process of setting up an environment to enable programming. Furthermore, I will introduce the art of programming, focusing directly on the practical aspects of analyzing real biological data and mining such datasets to design further hypothesis-driven studies. Although this chapter is aimed at students in Biology with limited experience in programming, the concept of programming for biological datasets should be helpful to those students in Computer Science as well.

3.2 Setting up an environment

Most bioinformaticians use either Linux or Mac as an environment for programming. However, such environments are often quite unfamiliar to students in Biology. Usually, students use a Windows environment for their daily tasks, which is in some areas less suitable for programming. However, there is a very convenient environment called 'Cygwin'. Cygwin (http://www.cygwin.com/) is a collection of tools to provide a Linux-like environment to run programs in Windows. As the Mac operating system is based on Unix, most commands are pre-installed to run programs. Therefore, Mac users can skip the next subsection. In the following, I will describe a way to install Cygwin to build an environment for programming in Windows.

From the web browser, open the URL: http://www.cygwin.com/. On the left hand side of the window, click on [Install Cygwin] to jump to a new page.

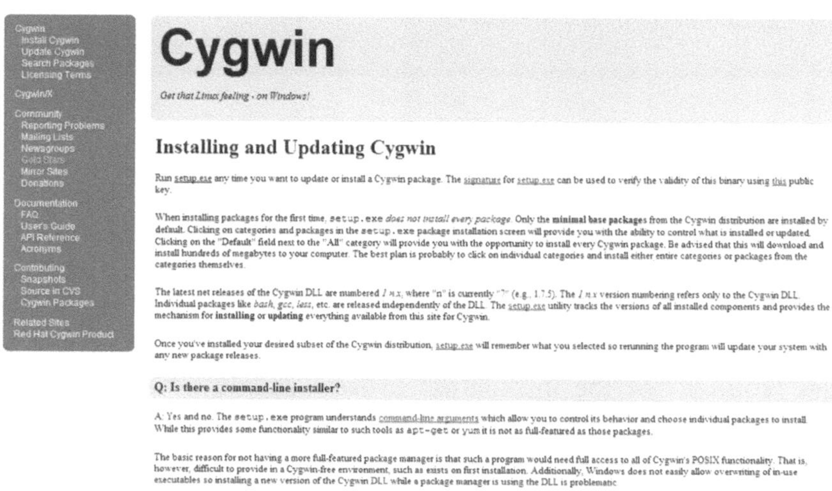

Click [setup.exe] to download the file for setting up Cygwin. Save it to your preferred location.

Double click on the saved [setup.exe] file ( ) to start the installation process. Click the [Next] button.

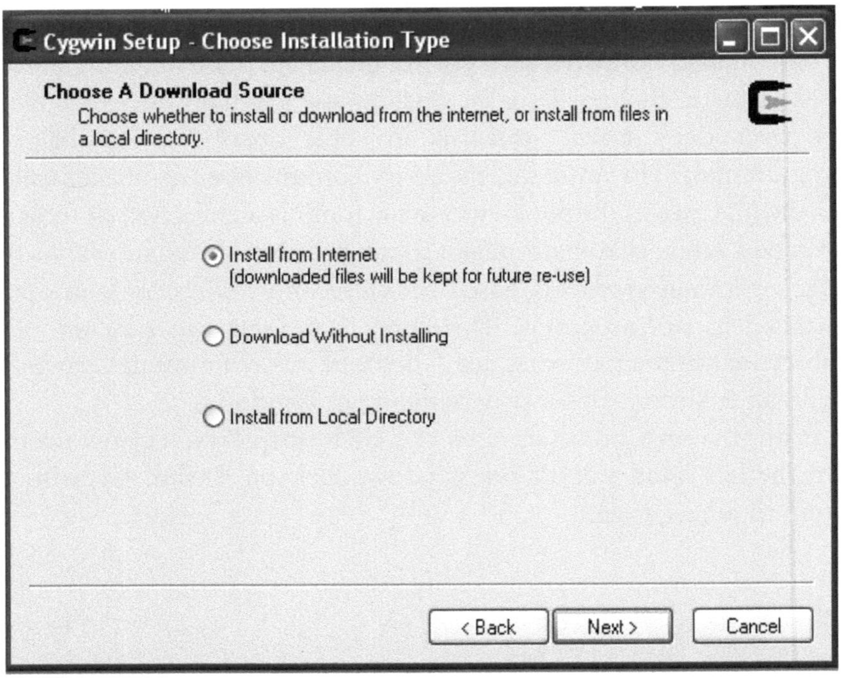

Select [Install from Internet] as shown above and click the [Next] button. As a default, the location of the installation for Cygwin is specified at '*C:cygwin*'. If you want to change to another location, specify it by clicking on the [Browse...] button. Click the [Next] button to continue. Specify the directory in which the downloaded files required for the installation process should be saved. If you have no special saving preferences, simply click on the [Next] button.

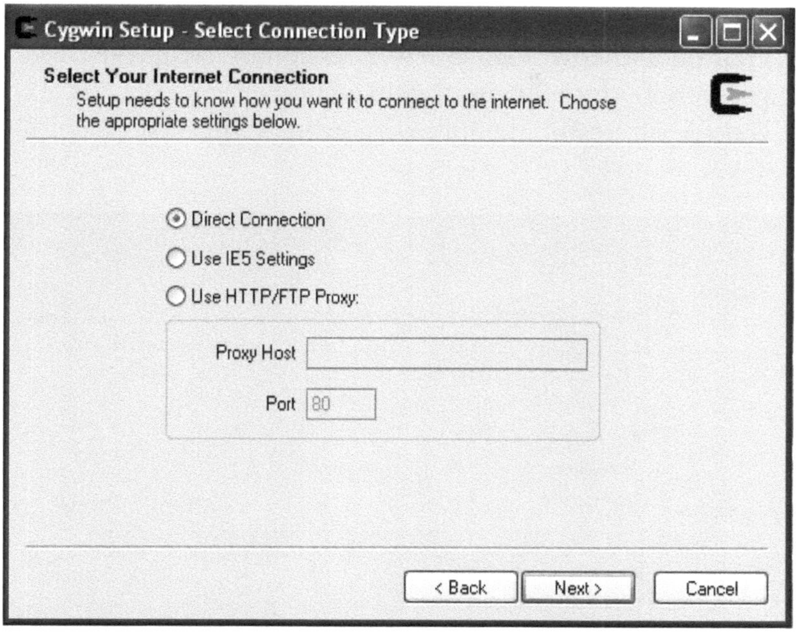

As a default, [Direct Connection] is selected. Simply click on the [Next] button to continue. Specify the site from which to download the files. It is recommended to select a site in your country to minimize the time required for the download. Press the [Next] button to continue.

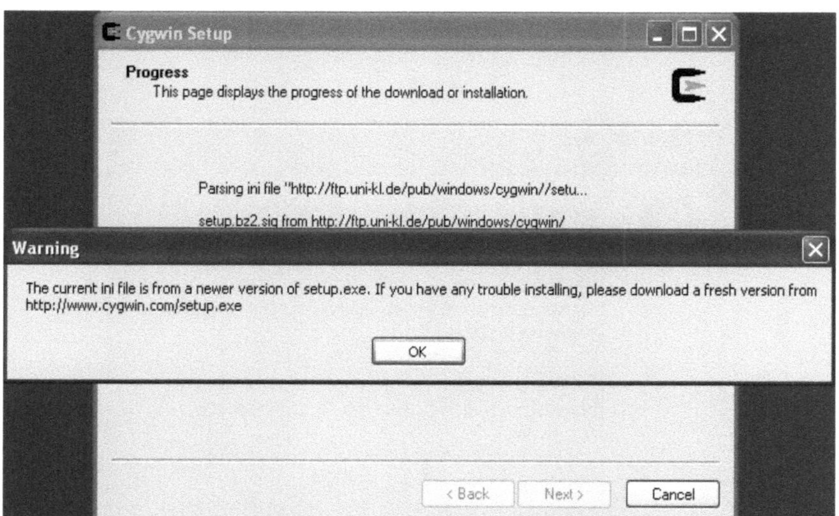

If you encounter the above warning message, simply click on the [OK] button and download a newer version of 'setup.exe' from the Cygwin website (http://www.cygwin.com/).

If you are successful, the following prompt appears:

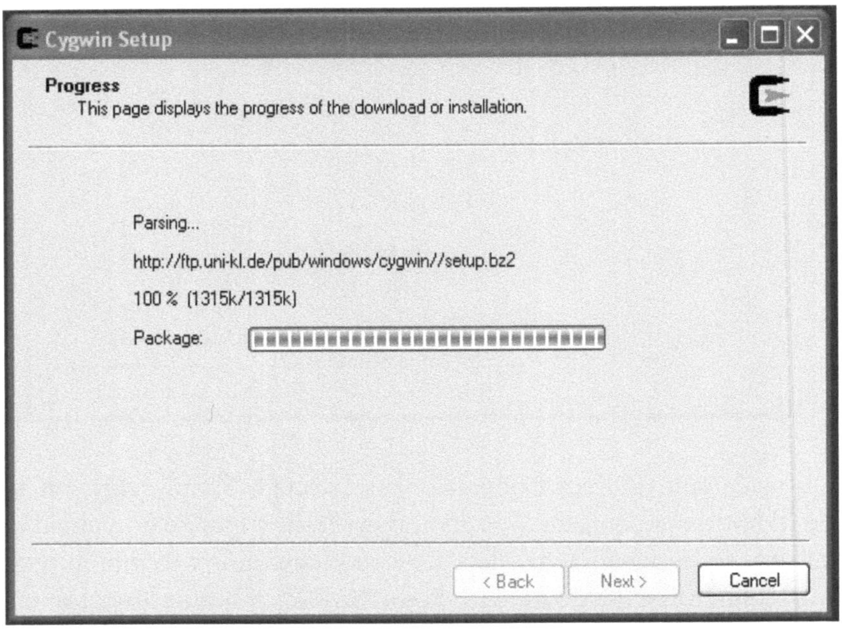

You might encounter a warning message as follows:

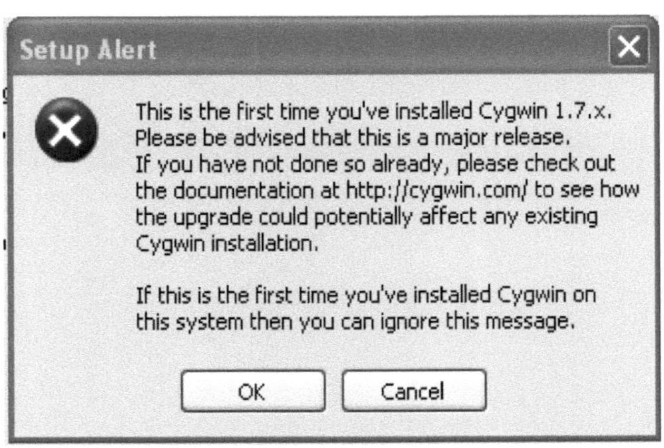

Press the [OK] button to continue. In the next step, you will select packages to be installed. To install all the packages necessary to run a Perl program, change the status of the 'Devel' packages to 'Install' from 'Default' by clicking on the word 'Default'.

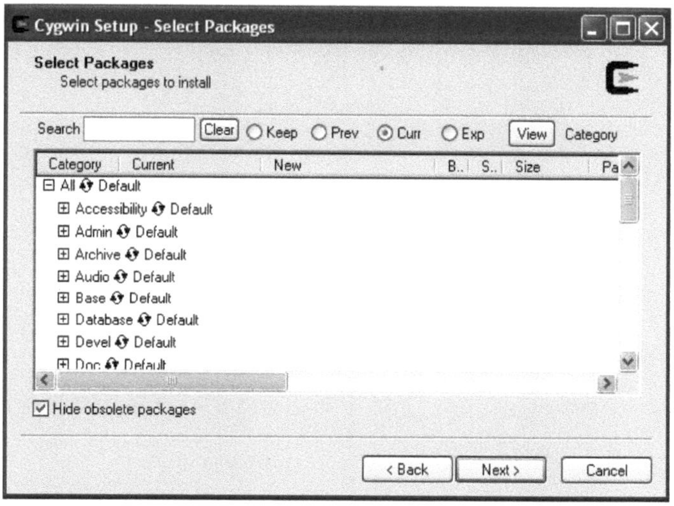

Press the [Next] button to continue. The package download starts immediately. When the following message appears, click the [Next] button to continue.

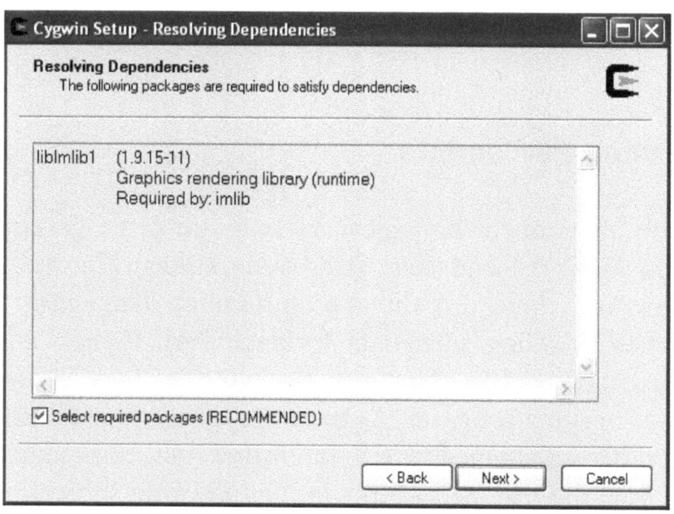

When the installation completes successfully, the following prompt appears.

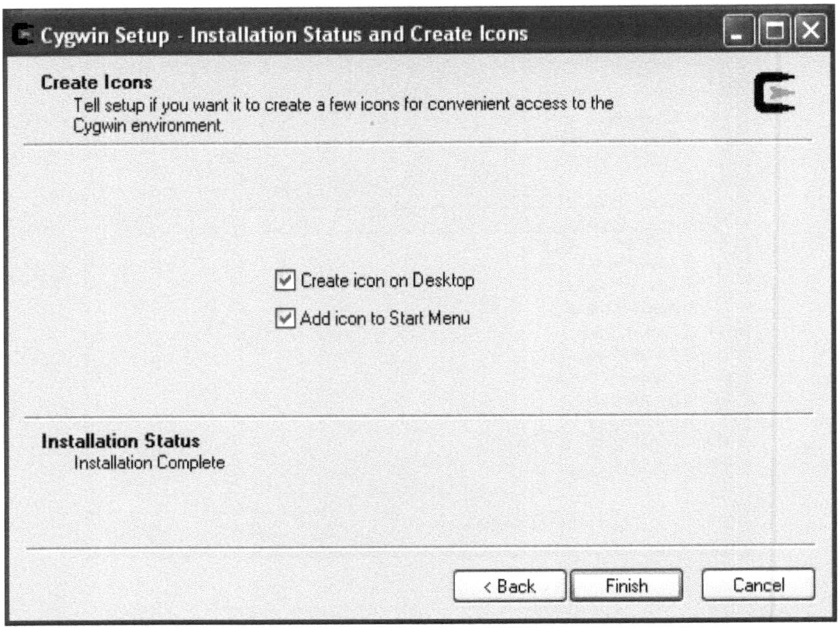

Check both boxes to create icons for convenient access to the Cygwin environment. By clicking on the [Finish] button, the installation process completes. You are now an owner of the Cygwin environment.

3.3 Writing Perl scripts

As already mentioned, biological datasets are often generated and stored as numbers and text (e.g. annotations). In the field of Bioinformatics, there are three programming languages that are useful when handling such data formats: Perl, Python and Ruby. These three languages are 'scripting languages' as they do not require a complier to run a program. As there are more users of Perl than of the other two languages in the Bioinformatics community, I will describe how to write Perl scripts in the following subsections.

3.3.1 What is PERL?

Perl was originally developed by Larry Wall (http://www.wall.org/~larry/) in 1987. He wrote a book called *Programming Perl* published by O'Reilly, a bible for Perl programmers. Perl is simple to use and powerful, and a community of users has developed and maintained the official 'Perl Programming Language' website (http://www.perl.org/). Because the Perl community is very strong, questions can be posed via various blog or websites listed at http://www.perl.org/community.html.

The reasons for using Perl in Bioinformatics are simple: it is easy to use and can easily handle text data (e.g. DNA sequences) through regular expressions (explained later in this chapter). As it is particularly popular for writing Common Gateway Interface (CGI) scripts, Perl is useful for web programming as well. By knowing how to program in Perl, you will open many doors in your research field.

3.3.2 Getting started

Although I will walk readers through how to program in Perl, it is important to remember the following famous Perl motto: 'There's more than one way to do it' (TMTOWTDI or TIMTOWTDI, pronounced 'Tim Toady'). Readers should modify the examples shown below for their purposes to explore the power and joy of using Perl.

If readers are using Cygwin, Perl is already pre-installed. You can check for the version of Perl by typing 'perl –v' in the console.

To get started, simply open any text editor (e.g. Notepad) and type:

```
#!/usr/local/bin/perl
print "Bioinformatics";
```

The first line '*#!/usr/local/bin/perl*' tells the operating system (Cygwin in our case) where to find the Perl interpreter to execute (or run) the program that you created. You do not need the first line in the Window's environment, even through Cygwin. However, not everyone is using the Cygwin/Windows environment. Even a beginner should be accustomed to using the first line to step up the ladder of Bioinformatics by using the Linux environment. One could modify this by adding '–w' at the end ('*#!/usr/local/bin/perl –w*') to allow for error messages to appear. This should help you to identify any errors that you may have made. You can also omit this and call this function at the time of running a program, which will be explained later. Technically, '#!' is called 'shebang notation' (also known as 'hash-bang' or 'sharp-bang'), which is used in the Unix environment to mark the start of a script. The second line is the actual job that you are giving the Perl interpreter to perform. Do not forget to end with a semicolon to tell the Perl interpreter that this is the end of the command line.

Save the above two sentences with a file name 'test1.pl' by creating a new folder on your computer. In the following, it is assumed that you have created a folder called 'test' under '/My Documents/'. '*.pl' specifies that this is a file with Perl commands.

Next, double click on the Cygwin icon on your desktop to start the Cygwin environment. In the Cygwin consol, type as follows to move to the folder that you have just created:

```
$ cd C:
$ cd Documents\ and\ Settings/
$ cd suchida/My\ Documents/test/
```

All the above commands can be shortened by simply pressing the [Tab] button on your keyboard, which allows you to select from the list of available folders. Note that 'suchida' is my user name on my computer.

From the directory where you saved the previous file, type '*perl test1.pl*', which instantaneously displays 'Bioinformatics'.

You can also type '*perl test1.pl –w*' to allow error messages to be displayed, if you have not done so in the program file ('*#!/usr/local/bin/perl –w*').

You can see that Perl is relatively simple to use, but powerful.

3.3.3 Some important things to know

Now that you can write and run a Perl program, I will run through the minimum set of variables and functions that one would need to know in order to process biological datasets when performing useful Bioinformatics analyses. This is like learning a new language. This part of the book may not be particularly interesting, but without it, the whole Perl programming will have many holes, which will hamper your further progress as a new Bioinformatician (as in any other occupation in the scientific field, there is no clear job description for a Bioinformatician, besides being able to write a program in a language such as Perl and to work with a Linux/Unix-based operating system). So please bear with me. I will make them as short as possible by highlighting the most important points. In the following subsections, *italics* are used for Perl programs whereas **bold** type is used for the output (result) of running the programs.

3.3.3.1 Scalar variables

A scalar variable contains only one element. This element can be a string ('I am a Bioinformatician'), number ('1', '2') or reference (e.g. subroutine, which is not described in this book). A string may contain symbols, letters, numbers or a combination of them. A scalar variable is prefixed with a dollar sign ($). The following are examples of scalar variables:

```
$codon = 3;
$dna = "deoxyribonucleic acid";
$atp = 29.85;
```

In the case of numbers, double quotations are not needed, but they must be used for a string.

3.3.3.2 Array variables

An array variable is composed of a list of values. Whereas a scalar variable can contain only one value, an array can store many. It is prefixed with an '@' symbol. For example:

```
@central_dogma = ("DNA", "RNA", "Protein");
```

In Perl, array indices start with 0. Square brackets are used to specify the index. So to refer to the first element of the array *@central_dogma*, you use *$central_dogma[0]*, which is '*DNA*'. Note that the dollar sign '$' must be used to refer to each element in an array.

As an array stores a list of variables in an ordered manner, there are several functions available in Perl to obtain a particular value from an array. The following functions are available (@central_dogma is used as an example: *@central_dogma = ("DNA", "RNA", "Protein");*):

- ***shift***: to take out the first element.

  $example1 = shift(@central_dogma);
  print "$example1\n";
  DNA

- ***pop***: to take out the last element.

  $example2 = pop(@central_dogma);
  print "$example2\n";
  Protein

- ***unshift***: to add an element to the array at the first position.

  unshift(@central_dogma, "Central Dogma");
  $example3 = shift(@central_dogma); # to take out the first element
  print "$example3\n";
  Central Dogma

- ***push***: to add an element to the array at the last position.

  push(@central_dogma, "Functions");
  $example4 = pop(@central_dogma); # to take out the last element
  print "$example4\n";
  Functions

- **splice**: to take out or add an element in any position of an array.
  - ***splice(ARRAY, OFFSET, LENGTH)***: to eliminate a part of an array.

  @central_dogma = ("DNA", "RNA", "Protein");
  @eliminated = splice(@central_dogma, 1, 1);
  print "@central_dogma\n";

DNA Protein

– *splice(ARRAY, OFFSET, LENGTH, LIST)*: to eliminate a part of an array and insert it in a new list.

```
@central_dogma = ("DNA", "RNA", "Protein");
@RNA = ("mRNA", "tRNA", "rRNA");
@eliminated = splice(@central_dogma, 1, 1, @RNA);
print "@central_dogma\n";
```

DNA mRNA tRNA rRNA Protein

3.3.3.3 Hash

A hash folds a complex list with both a key and a value part for each element of the list. It is prefixed with the per cent symbol '%'. For example:

Hash Name		**Key**	**Value**
%central_dogma	=(	'DNA',	'deoxyribonucleic acid',
		'RNA',	'ribonucleic acid',
		'Protein',	'amino acids'
	);		

Another way to define a hash would be as follows:

```
%central_dogma   =(  DNA       => 'deoxyribonucleic acid',
                     RNA       => 'ribonucleic acid',
                     Protein   => 'amino acids'
                 );
```

The => operator is a synonym for a comma in the previous example. It automatically quotes the left side of the argument as shown in the above example.

To print out the elements of a hash, the easiest way is to use a foreach loop. For example:

```
foreach $element (keys (%central_dogma)){
  print "$element\t$central_dogma{$element}\n";
}
```

DNA	**deoxyribonucleic acid**
RNA	**ribonucleic acid**
Protein	**amino acids**

Because many new functions are used above, I will explain them one by one. A foreach loop is a way to repeatedly run commands within "{ }" until a certain condition is met: in the above case, as long as there are elements in *%central_dogma*. Each time the loop runs, it copies an element in the hash and puts it into a scalar variable named *$element*. In the above example, the elements will be displayed in random order. For display in an ordered fashion, the *sort* function must be used. For example, in order to sort by keys:

```
foreach $element (sort keys (%central_dogma)){
  print "$element\t$central_dogma{$element}\n";
}
```

DNA	**deoxyribonucleic acid**
Protein	**amino acids**
RNA	**ribonucleic acid**

Note how the order of **RNA** and **Protein** has been changed. In order to sort by values:

```
foreach $element (sort {$hash{$a} <=> $hash{$b}} keys
(%central_dogma)){
  print "$element\t$central_dogma{$element}\n";
}
```

Protein	**amino acids**
DNA	**deoxyribonucleic acid**
RNA	**ribonucleic acid**

A hash is a convenient way to annotate DNA, RNA or proteins.
There are a number of functions available for hash:

- ***delete $hash{$key}***: to delete the specified key and value pair and to return the deleted value.
- ***exists $hash{$key}***: to return true if the specified key exists in the hash.
- ***keys %hash***: to return a list of keys for the specified hash. (Its use has been shown in the above example.)
- ***values %hash***: to return a list of values for the specified hash.
- ***scalar %hash***: to return true if the hash has elements defined. This function is used to check if a hash is empty or not.

3.3.3.4 File handling

Perl is equipped with fairly comprehensive functionalities for file input and output. For example:

```
$file = "biology.txt";
$new = "revised.txt";
open (IN, $file) or die "$! ($file)";
open (NEW, ">", $new) or die "$! ($new)";
$_ = <IN>;
while (defined ($line = <IN>)) {
  print NEW "$line";
}
close (NEW);
close (IN);
```

The first two lines indicate the scalar variables '*$file*' and '*$new*', which store the file names 'biology.txt' and 'revised.txt', respectively. The third line is an operation for file handling. The **open** function opens 'biology.txt'. The first parameter shown here is a file handle. A file handle is a variable dealing with a file, which allows Perl to refer to the file in future operations. As a rule, it should be written in capital letters, '*IN*' in the above example. Separated from the first by a comma, the second parameter is an expression denoting the file, which is '*$file*' in this example. '*die "$! ($file)"*;' indicates that if the specified file (i.e. "*biology.txt*") does not exist, then the script will die with an error message stored in the special variable '$!' and the path of the missing file as stored in '*$file*'. The fourth line is similar to the third line, except the second parameter, which is '">", "*$new*"'. The operand '>' highlights the file to be written, which is *$new* in this case. The fifth line is to read the file line by line and store it in a special variable '$_', which is a default input that can be used for global purposes. The sixth line is a while loop. The *<IN>* operator returns all the unread lines of the file one by one in scalar context. The seventh line uses a print function to print the content of the *$line* variable into the *$new* file using the *NEW* file handle. The program above ends by closing the files in reverse order in which they were opened.

3.3.3.5 Regular expressions

One of the hallmarks of Perl is regular expressions. These can be used to manipulate any types of string data. The most common use of regular expressions is pattern matching. This is not a proprietary function of Perl, and has been used in a variety of applications, including World Wide Web search engines (e.g. Google). There are many things that can be done with regular expressions, and I will highlight only a few. For interested readers, I highly recommend the book *Mastering Regular Expressions* by Jeffrey E. F. Friedl (Oreilly & Associates Inc., ISBN: 0-596-52812-4). Through this book, you can really 'master' regular expressions to a professional level.

3.3.3.5.1 Pattern matching

Pattern matching in Perl is useful because most gene annotations are in text form (e.g. the official gene name of '*Myh6*' is 'myosin, heavy polypeptide 6, cardiac muscle, alpha'), and also biological databases are available in raw text files. This tool is particularly helpful if you are given microarray data with over 40,000 probes with annotations as gene names and symbols and are trying to find a gene of interest (e.g. *Myh6*). You can simply read in the file and use the pattern matching command using an equal sign '=' and a tilde '~' followed by your pattern of interest enclosed within two slashes '/' called an operand (e.g. '=~ /*Myh6*/'). Note that as a default, pattern matching is case sensitive. In other words, the above example of pattern matching does not match to 'myh6'. This can be overcome by using a modifier called '*i*' (e.g. '=~ /*Myh6*/*i*'). There are other modifiers available:

- *m*: to perform a matching for multiple lines
- *o*: to compile a pattern only once to optimize the search
- *s*: to treat the whole string as one line
- *x*: to allow white space and comments within the regular expression search

In addition, if you want to match words in specific locations and patterns, the following metacharacters are available:

- *^*: to match at the beginning of a string
- *$*: to match at the end of a string.
- *(...)*: to treat a part of a string as a unit

The above explanation is by no means complete. There are very many more functions available through regular expressions, which you will need to explore by yourself.

3.3.3.5.2 Match and replace

If you use an operator called 'substitution', not only can you match a word or text in a search but you can also replace it with others. To use this operator, simply add another slash called 'replacement operand' followed by the word or text to be added.

3.3.4 Some examples

Let us now look at some specific examples to become accustomed to programming in Perl. I need to stress again that Perl is not difficult! It is much easier than learning a foreign language. Do not be frightened by making a mistake. You can always correct the programming code to run the Perl interpreter again to see the results that you expected.

3.3.4.1 Example 1: transcription

The central dogma of Biology is that DNA is transcribed to RNA, which is translated to proteins. Although both DNA and RNA are nucleic acids, they differ as follows:

- The sugar present in a DNA molecule is deoxyribose, whereas that of RNA is ribose.
- DNA forms a double strand, whereas RNA exists as a single strand.

- DNA is found in the cell nucleus, whereas RNA is found in the nucleus and/or cytoplasm.

Another difference lies in the bases: in DNA, adenine (A), thymine (T), guanine (G) and cytosine (C). In the case of RNA, thymine is replaced with uracil (U). This difference is evident when a transcription takes place. This process of transcription can be copied on the computer by using the following Perl script:

```
#!/usr/local/bin/perl -w
# This is the input DNA sequence
$DNA = "ATGCGCTTAAG";
# First, initialize the RNA sequence
$RNA = "";
# Copy the DNA sequence
$RNA = $DNA;
# Substitute all T's with U's
$RNA =~ s/T/U/g;
# Print out the result
print "DNA:\t$DNA\n";
print "RNA:\t$RNA\n";
```

When you run the above program in the Cygwin console, the following result should appear:

```
DNA:   ATGCGCTTAAG
RNA:   AUGCGCUUAAG
```

Let us go through the above script line by line. The first line is the magic word in the Perl programming, which you always need to use to find the Perl interpreter. To make debugging of the program easier, I inserted '-w' for error messages. The second line starts with a sharp '#', which indicates that this is a comment line, and will not be processed by the Perl interpreter as a part of a job that it has to execute. The third line is the input DNA sequence. As written in the fourth line as a comment, a variable must be initialized as in the fifth line. The reason behind this is that the variable that is specified above (here '$RNA') may have been used before, not in this program but by others (or perhaps yourself) using the same environment (i.e. Cygwin). If you do not initialize the variable, you might obtain a value that you had not intended to see when your program is executed. To avoid this potential problem, I recommend initializing any variables (this holds true for an array and hash) in each program. To convert the sequence, first copy the DNA sequence as in the seventh line. As stated in the eighth line as a comment, using a substitute operator together with a pattern modifier called 'global' ('g'), as described in the previous subsection, the copied DNA sequence will be converted to the RNA sequence by substituting thymine (T) by uracil (U). The program ends by printing out the result. Note that I used a tab ('\t') to align the output result for easier readout. So far so good, I hope. Let's move on to another example.

3.3.4.2 Example 2: DNA complementarity

One of the first things that you learned in Biology is that DNA usually exists as a pair of molecules that are held tightly together in a double helix. To form a double helix, the two strands must be complementary to each other. Using Perl, you can convert a single-stranded DNA sequence into a complementary sequence. For example, if you are given 'ATGC' as a sequence, you can write the following to make a complementary sequence:

```
#!/usr/local/bin/perl -w
# Make a reverse complement sequence
# This is the input sequence
$input_seq = "ATGC";
# To convert to a complementary sequence,
# first, a variable called '$output_seq' must be initialized
$output_seq = "";
# Copy the whole sequence
$output_seq = $input_seq;
# Convert the sequence
$output_seq =~ tr/ATGC/TACG/;
# Print out the result
print "Original sequence:\t$input_seq\n";
print "Complementary sequence:\t$output_seq\n";
```

The output from the above example program should look as below on your screen:

Original sequence: ATGC
Complementary sequence: TACG

Note that in this example, the tr operator to translate a set of characters (i.e. "ATGC") into new characters was used. Compared with the substitute command, the tr operator converts all the characters at once. Let us try a more challenging example.

3.3.4.3 Example 3: translation

After DNA is transcribed into RNA, RNA is translated into a protein by ribosomes. There is a set of rules called 'genetic code' by which three ribonucleic acids are converted to one amino acid. According to current understanding, there are different types of genetic codes available based on organisms and organelles (e.g. mitochondria). The list of translation tables can be found at: http://www.ncbi.nlm.nih.gov/Taxonomy/Utils/wprintgc.cgi. To make things easier, the following example program uses the standard genetic code (http://www.ncbi.nlm.nih.gov/Taxonomy/taxonomyhome.html/index.cgi?chapter=tgencodes#SG1):

```
#!/usr/local/bin/perl -w
# Translate a mRNA sequence into amino acids
# This is the input sequence
$input_seq = "ATGACGGACGCCCAGATGGCTGACTAA";
# Use a hash table to make a translation table.
# It will be easier to access the site mentioned above
# to copy and paste the translation table
# '*' represents a stop codon.
%table = ();
%table = (
    "TTT" => "F",
    "TCT" => "S",
    "TAT" => "Y",
    "TGT" => "C",
    "TTC" => "F",
```

```
"TCC" => "S",
"TAC" => "Y",
"TGC" => "C",
"TTA" => "L",
"TCA" => "S",
"TAA" => "*",
"TGA" => "*",
"TTG" => "L",
"TCG" => "S",
"TAG" => "*",
"TGG" => "W",
"CTT" => "L",
"CCT" => "P",
"CAT" => "H",
"CGT" => "R",
"CTC" => "L",
"CCC" => "P",
"CAC" => "H",
"CGC" => "R",
"CTA" => "L",
"CCA" => "P",
"CAA" => "Q",
"CGA" => "R",
"CTG" => "L",
"CCG" => "P",
"CAG" => "Q",
"CGG" => "R",
"ATT" => "I",
"ACT" => "T",
"AAT" => "N",
"AGT" => "S",
"ATC" => "I",
"ACC" => "T",
"AAC" => "N",
"AGC" => "S",
"ATA" => "I",
"ACA" => "T",
```

```
    "AAA" => "K",
    "AGA" => "R",
    "ATG" => "M", # Start codon
    "ACG" => "T",
    "AAG" => "K",
    "AGG" => "R",
    "GTT" => "V",
    "GCT" => "A",
    "GAT" => "D",
    "GGT" => "G",
    "GTC" => "V",
    "GCC" => "A",
    "GAC" => "D",
    "GGC" => "G",
    "GTA" => "V",
    "GCA" => "A",
    "GAA" => "E",
    "GGA" => "G",
    "GTG" => "V",
    "GCG" => "A",
    "GAG" => "E",
    "GGG" => "G"
);
# divide the input sequence into codons
# first put into an array
@RNA = ();
@RNA = split(/(\w\w\w)/, $input_seq);
# initialize
$RNA = " ";
$protein = " ";
# take out each codon and convert
for(my $i=0; $i<(length($input_seq)-2); $i+=3) {
    $codon = substr($input_seq,$i,3);
    $RNA .= $codon . " ";
    $protein .= " " . $table{$codon};
}
# print out the result
```

```
print "mRNA: $RNA\n";
print "Protein:$protein\n";
```

The output from the above example program should look like this:

```
mRNA:    ATG ACG GAC GCC CAG ATG GCT GAC TAA
Protein:  M   T   D   A   Q   M   A   D   *
```

The outputs are aligned by spaces so that the input mRNA sequence can be compared easily with the corresponding amino acid codon.

3.3.4.4 Example 4: merging datasets

In the above three examples, only one input (i.e. sequence) is used for each program. A computer can handle much more information in milliseconds. Therefore, in this example, I will introduce you to file functions. This set of functions is very useful when you want to merge two or more large tables. To start this example, please access the following site at the Mouse Genome Informatics (MGI): ftp://ftp.informatics.jax.org/pub/reports/index.html. Download the following two files: 'MGI_Coordinate.rpt' and 'MRK_Reference.rpt'. You will obtain a list of PubMed IDs (PMIDs) for each gene.

First, open the 'MRK_Reference.rpt' dataset and make a hash table called '%table' with an MGI Marker Accession ID as a key and a list of PMIDs as a value.

```
#!/usr/local/bin/perl
# Initialize
$file1 = "MRK_Reference.rpt";
%table = ();
# open $file1 and make %table
open (F1, $file1) or die "$file1";
while (defined ($f1 = <F1>)){
    # This step is necessary to remove unwanted symbol
    # created by the Windows.
    $f1 =~ s/\r\n/\n/;
    # Remove the new line
    chomp ($f1);
    # Initialize
```

```
        @data = ();
        @data = split(/\t/, $f1);
        # Make %table
        $table{$data[0]} = $data[4];
    }
```

To check that you have placed all the necessary information into a hash table, write as follows:

```
    foreach $i (sort keys (%table)){
      print "$i\t$table{$i}\n";
    }
```

To run the program above, save the file with a name like 'file_handle.pl' and run with the *less* command to view the result one screen at a time. Type in 'perl file_handle.pl | less' in the command prompt of Cygwin.

```
MGI:100000  8211130|7920646|8600386|10431246|9503013
MGI:100001  8211130|7920646|8600386|10431246|15020476
MGI:100002  8211130|7920646|8630493|8600386|9880679|10431246
            ...
```

As you can see from the output result, first, the MGI Marker Accession ID is displayed followed by a tab and the list of PMIDs separated by a vertical bar ('|') (also known as a 'pipe' symbol).

Next, open the 'MGI_Coordinate.rpt' dataset to give an Entrez Gene ID (GI) used for the list of corresponding PMIDs. This time, the output will be saved to a new file called 'Result_file_handle.txt'.

```
    # Initialize
    $file2 = "MGI_Coordinate.rpt";
    $new = "Result_file_handle.txt";
    # open $file2
    open (F2, $file2) or die "$file2";
    open (NEW, "> $new") or die "$new";
```

```
# Skip the header line.
$_ = <F2>;
# make a new header line
print NEW "GI\tPMID\n";
while (defined ($f2 = <F2>)){
  $f2 =~ s/\r\n/\n/;
  chomp ($f2);
  # Initialize
  @data = ();
  @data = split(/\t/, $f2);
  # Match up
  if ($table{$data[0]}){
    print NEW "$data[10]\t$table{$data[0]}\n";
  }
}
close (NEW);
close (F2);
```

The output result is saved as a file, which you can open in Microsoft Excel. You will notice that some GIs are indicated as 'null'. This means that the corresponding Entrez Gene ID is missing for an MGI Marker Accession ID.

If you would like to do the same with other organisms, please access the Entrez Gene FTP site (ftp://ftp.ncbi.nih.gov/gene/DATA/) and download the following two files: 'gene2accession.gz' and 'gene2pubmed.gz'. You can expand these gzip files via Cygwin. Type in the following in the directory where the files are saved: 'gzip -d *.gz'. Note that the two files are very big and will take a long time to process.

3.3.5 Don't reinvent the wheel

Perl is the most often used program to process biological data, and as such there are scripts available via the internet. The Bioperl Project (http://www.bioperl.org/wiki/Main_Page) is designed to meet such demands. If you are interested in using Bioperl, please refer to the following URL for installation: http://www.bioperl.org/wiki/Installing_Bioperl_on_Windows#Bioperl_in_Cygwin. Besides Bioperl, I can also recommend a web page set up by Techcuriosity.com for the Bioinformatics Perl Script Library (http://www.techcuriosity.com/resources/bioinformatics/scripts_library.php). Furthermore, the Comprehensive Perl Archive Network (CPAN) (http://www.cpan.org/) and Google Code (http://code.google.com/) also contain useful scripts. If you make useful scripts, please try to contribute to the above-mentioned projects.

3.4 Conclusion

Computers are not designed just to check emails and read articles. Furthermore, biological primary and secondary databases are not built just to be searched through pre-built query interfaces. The two are conveniently available to you in most laboratory settings to be explored and maximized. I hope I have convinced you that Perl is really easy to use. In fewer than 10 lines, you can do most of the jobs that biologists previously spent days and even months on (e.g. checking to see if two sequences from different species match each other). In the next chapter, I will introduce how we used Perl programs to build our knowledge database called 'C-It' (http://C-It.mpi-bn.mpg.de) (Gellert *et al.*, 2010) to screen for evolutionarily conserved, tissue-enriched, uncharacterized genes.

References

Blekhman R, Marioni JC, Zumbo P, Stephens M, Gilad Y. (2010) Sex-specific and lineage-specific alternative splicing in primates. *Genome Res* 20(2): 180–9.

Gellert P, Jenniches K, Braun T, Uchida S (2010) C-It: a knowledge database for tissue-enriched genes. *Bioinformatics* 26(18): 2328–33.

Stetefeld J, Ruegg MA. (2005) Structural and functional diversity generated by alternative mRNA splicing. *Trends Biochem Sci* 30(9): 515–21.

Wang ET, Sandberg R, Luo S, *et al.* (2008) Alternative isoform regulation in human tissue transcriptomes. *Nature* 456(7221): 470–6.

4

Case study: C-It, knowledge database for screening evolutionarily conserved, tissue-enriched, uncharacterized genes

Abstract: The tremendous speed of experimental data generation in recent years has posed numerous problems in the field of Bioinformatics. One of the most pressing problems is extracting knowledge from all these biological data. In this chapter, I describe our approach to screen such datasets *in silico* via our 'C-It' tool to make predictions to annotate evolutionarily conserved, tissue-enriched, uncharacterized genes.

Keywords: C-It, evolution, *in silico* screening, tissue, uncharacterized genes

4.1 Introduction

As I have already noted, the amount of biological data is now in the order of terabytes. The need to process such a huge amount of data is a daunting task for biologists and bioinformaticians. Although this chapter focuses on *in silico* screening and its applications, the examples that I give from our studies and products are based solely on data-mining currently available datasets to narrow down the list of unknown genes so that biologists can perform further experiments to understand their functions. Once their functions are elucidated, bioinformaticians can utilize such datasets to further refine the list of unknown genes. Ultimately, we would have a complete list of genes and their functions.

4.2 Concept of *in silico* screening

Through the development of high-throughput methods (e.g. microarrays, deep sequencers and mass spectrometry), an overwhelming amount of biological data is generated every day around the world. It is now a standard requirement that once a manuscript is accepted by a journal, the raw and/or analyzed data should be disclosed to the scientific community through public databases and/or the researchers' own websites. Such databases are listed in Chapter 2. Although the researchers who collected such datasets (mainly biologists and clinicians) had a certain hypothesis in mind when generating them, the data are merely a collection of data points (number of cells, phenotypic counts, expression values, images of morphological changes, etc.) at a particular time when the experiments took place. To confirm such observations, the researchers often conduct experiments under different conditions and/or at different time points and repeat the whole set of experiments to minimize variations. 'Biological replicates' is the term used when different biological materials (e.g. cells cultured on different days) are employed, whereas 'technical replicates' means that the same set of materials (e.g. RNA) is used but is processed on different days or in a different way. In essence, such data can be considered as a picture; and depending on how you look at it, this picture can be interpreted differently. This is the point to keep in mind for *in silico* screening. If you are working in purely 'dry' lab conditions (without any biological experiments), it is impossible to generate such data. Also, conducting large-scale biological experiments costs a lot of money. However, data can be analyzed in multiple dimensions, which is the classical thinking in the field of Bioinformatics. We can thus save time and money when collecting biological data by performing experiments. Nothing is free (e.g. reagents) in real life, certainly not the lives of animals. We live with the concept of 'recycling' to give a 'greener' society. By re-using the datasets generated by others, we are actually going 'green' and intellectually 'recycling' resources. With *in silico* screening, hypothesis-driven research can be conducted with less money and time by 'recycling' datasets.

4.2.1 Data collection

After making a hypothesis (e.g. 'Are there tissue-specific or tissue-enriched genes without functional annotations?'), the *in silico* screening process starts by collecting datasets that are necessary to test the hypothesis. The beauty of *in silico* screening is that you do not have to design experiments, order and buy reagents, culture cells or sacrifice an appropriate model animal. The only thing you need is a computer and a connection to the internet. Through the emergence of various experimental techniques and the places (e.g. databases) in which these generated datasets and their analyzed results are deposited, we have more than enough publicly available information that can be downloaded via the internet. Although it is scientifically interesting and important to work with primary data (unpublished), one could still try to conduct an *in silico* screening without such primary datasets.

Where should we go to obtain the data to test our hypothesis? Some useful databases and datasets are listed in Chapter 2, all of which can be accessed via the internet. Download them and try your programming skills.

4.2.2 Development of algorithms

After the appropriate datasets are downloaded, you need to extract the necessary information from them. Perl is very useful in this regard owing to its regular expression function. Pattern matching is the key here. For example, you can take out annotations about genes by downloading the 'gene_info' dataset from the Entrez Gene database (ftp://ftp.ncbi.nih.gov/gene/DATA/). From this dataset, you can obtain

GeneIDs, gene symbols, gene names, etc., by simply using an array and the split function to take out the information you are looking for (please refer to Chapter 3 for details about Perl programming).

The next thing you need to think about is how to connect different datasets in a manageable manner and how to develop an algorithm by which you can process the merged datasets and extract knowledge. Put simply, an 'algorithm' is a way to solve a problem in a finite number of steps. A common mistake is that people think that an algorithm must be mathematically based on very complicated formulas. In fact, as in any other fields (e.g. manufacturing of cars), the ultimate goal of an algorithm is to transfer the knowledge that experts in a certain field have to model their thinking and skills so that a computer can take over their task. In the case of Biology, experimental biologists are the experts. Often, such knowledge is based on experimental experience. Therefore, although it is trendy to come up with mathematical formulas, the most important thing to remember when developing an algorithm is the transfer of knowledge of experienced biologists; in my opinion, the ultimate aim of bioinformaticians should be to develop such algorithms. After all, *in silico* screening is nothing but a prediction, which needs to be tested and validated by biological experiments.

4.2.3 Visualization of analyzed results

Using an algorithm allows you to process any datasets that you have gathered through various databases and websites. Once this process is finished, you have analyzed the data and in most cases you will have a table of results. These results can be annotations of genes, names of protein domains, chromosomal positions, expression values from microarray experiments, etc. Depending on the size of the primary datasets, such results can be very extensive. For example, in a typical microarray experiment, you will obtain more than 10,000 probe sets, including expression values (e.g. normalized ratio). Unless you really like going over the table row-by-row and column-by-column in Microsoft Excel, it is sensible to visualize such datasets in some way. This could be some kind of graph (Venn diagram), viewers (e.g. UCSC Genome Viewer), etc. This area of study has attracted tremendous attention by researchers in recent years. Therefore, it is not necessary to build one for yourself. You can simply search for such software products through the internet to find the one that fits your requirements. One such place to look is the '.NET Bio' (http://bio.codeplex.com/).

4.3 Case study: C-It

Biology deals with the dynamics of largely irreversible processes (e.g. enzyme activities) and the relation of existential dependence in time and space (e.g. cell differentiation) in much more complicated terms than in the fields of Physics and Chemistry (Hutten, 1960). Thus, generalizations regarding biological processes are hard to achieve; this relies mostly on experiences that biologists have had and passed down from generation to generation in the form of teaching and lab rotations. Since the appearance of Molecular Biology, many kits and products have been sold by various companies, so even without understanding the underlying concept upon which each experiment was based, experiments could be performed to obtain results to test hypotheses. These technical parts are somewhat generalized, but the concept of uncovering (that is, the laws of nature) is not something that could be done in one experiment or in overnight reactions. As explained above, the ultimate goal of bioinformaticians is to transfer the knowledge of biologists into algorithms (so that computer programs can be written) to predict the 'unknowns'. This is exactly what *in silico* screening is based on.

4.3.1 UniGene

In April 2007, a month after finishing my PhD, Dr André Schneider, a group leader at the Max Planck Institute of Heart and Lung Research (Bad Nauheim, Germany), approached me with his way of reading the UniGene database. As an experienced developmental biologist, Dr Schneider had a distinctive way of interpreting the EST (expressed sequence tag) profiles of the UniGene database (see 2.3.1.1.2 UniGene for details). The following are a set of his heuristic rules that we applied to screen for 'evolutionarily conserved (four organisms: chicken, mouse, rat and human), heart-enriched genes' (Uchida *et al.*, 2009):

- Select genes whose rank for 'heart' is under the top 20% across homologs.

The above rule resulted in the selection of heart-specific genes as well as ubiquitously expressed genes. To eliminate the latter, the following further selection rules were developed:

- Further Selection Rule #1: Expressed in less than 50% of total tissues.
- Further Selection Rule #2: Sum of Gene EST must be less than 400.
- Further Selection Rule #3: At least 25% of Gene EST must be expressed in the target tissue.

The above set of rules was used to develop a simple algorithm (DGSA = Database-dependent Gene Selection and Analysis). To apply DGSA, Perl programs were written to screen using mouse genes as a reference:

Step 1 Create a list of homologs based on the HomoloGene group ID (HID) from five species: *Homo sapiens* (human) (Taxonomy ID = 9606); *Rattus norvegicus* (rat) (10116); *Mus musculus* (mouse) (10090); *Gallus gallus* (chicken) (9031); and *Danio rerio* (zebrafish) (7955). The dataset was downloaded from the HomoloGene FTP site (ftp://ftp.ncbi.nih.gov/pub/HomoloGene/).

Step 2 Find the UniGene IDs matching to GIs using the 'gene2unigene' dataset from Entrez Gene (ftp://ftp.ncbi.nih.gov/gene/DATA/).

Step 3 Obtain the UniGene profiles for each GI from Step 2 for each organism using the datasets in ftp://ftp.ncbi.nih.gov/repository/UniGene/.

Step 4 Rank the tissues according to the UniGene profiles and apply DGSA to select tissue-enriched genes for each organism.

Step 5 Attach PMIDs by excluding those articles with more than 100 genes using the 'gene2pubmed' dataset (ftp://ftp.ncbi.nih.gov/gene/DATA/).

The validity of our *in silico* screening was tested by a reverse transcription polymerase chain reaction (RT-PCR) using 15 mouse adult organs, whole-mount *in situ* hybridization (WISH) using 9.5 and 10.5 days post-coitum mouse embryos and an antisense morpholino-based knockdown in zebrafish embryos (Sumanas and Larson, 2002). The results can be freely accessed at: http://www.ncbi.nlm.nih.gov/pmc/articles/PMC2657154/ (Uchida *et al.*, 2009). As it was possible to validate the expression patterns and observe phenotypes in the target organ (i.e. heart), our *in silico* screening for evolutionarily conserved, tissue-enriched, uncharacterized genes can convert biological information into knowledge.

Based on the *in silico* screening of the UniGene profiles using our DGSA algorithm, it is now possible to search for a set of evolutionarily conserved, tissue-enriched, uncharacterized genes. To further refine the list of potentially interesting genes, other datasets as listed below were incorporated to build a knowledge database called 'C-It' (http://C-It.mpi-bn.mpg.de) (Gellert *et al.*, 2010):

- SymAtlas
- GeneChip Exon 1.0 ST Arrays (Exon Arrays)
- Mouse Atlas of Gene Expression

In the following subsections, each dataset is described in detail.

4.3.2 SymAtlas

At present, DNA microarrays (also known as gene chips, gene arrays) represent the most commonly used mRNA profiling technique, and have proved to be an efficient tool for obtaining gene expression profiles at particular time points. In the case of the GeneChip Gene 1.0 ST Array (Affymetrix, Inc.), which is commonly known as 'gene arrays', only 50 ng of total mRNA is needed to perform an array experiment. In most cases, 2–3 days are required to generate the primary dataset. Through standardization of the protocols, systematic biases (noises) arising from microarrays [e.g. operator bias (Quackenbush, 2002; Uchida *et al.*, 2005)] have been minimized, which was supported by studies of the MicroArray Quality Control (MAQC) consortium (MAQC Consortium, 2006).

The GNF SymAtlas2 project performed microarray experiments using custom Affymetrix GNF1M arrays to provide a reference set of gene expression profiles of various human and mouse tissues (Su *et al.*, 2002). This is now integrated into 'BioGPS', which is a comprehensive gene annotation portal (http://biogps.org/) (Wu *et al.*, 2009). All the raw data (Affymetrix's CEL files) can be downloaded from the Gene Expression Omnibus (GEO) (http://www.ncbi.nlm.nih.gov/geo/query/acc.cgi?acc=GSE1133). For example, in the case of mouse, there are microarray datasets of 61 tissues with two samples per tissue available. For our purpose, all data were normalized using the R statistical environment (www.r-project.com) and the Bioconductor package 'affy' by the robust multi-chip average (RMA) method (Gautier *et al.*, 2004). To obtain differentially expressed genes, an analysis of variance (ANOVA) statistical test was performed: that is, the mean expression of a gene for a target tissue against the mean expression value in all other tissues to give a *P*-value for the enrichment in the target tissue. This test was applied for every probe set on a microarray chip.

4.3.3 Exon arrays

Alternative splicing is a post-translational mechanism to diversify the number of proteins to be produced from pre-mRNA sequences. With the rise of proteomics methods (e.g. 2-D gels and mass spectrometry), alternative splicing is considered to be one reason for the discrepancies between transcriptomics and proteomics data (Bitton *et al.*, 2008). It is predicted that more than 70% of the human genes undergo alternative splicing (Johnson *et al.*, 2003), and the most recent reports estimated this number to be more than 90%

(Pan *et al.*, 2008; Wang *et al.*, 2008). Due to very frequent splicing events, mutations in alternative splicing machineries result in various diseases (Faustino and Cooper, 2003; Wang and Cooper, 2007). It is now widely accepted that most of the alternative splicing events are controlled in a tissue-specific manner (Blencowe, 2006). There are numerous reports of genome-wide events of tissue-specific alternative splicing (Modrek *et al.*, 2001; Xu *et al.*, 2002; Johnson *et al.*, 2003; Yeo *et al.*, 2004), mostly in brain and testes where there are high incidents of splicing compared with other tissues, as well as cancer-specific splicing (Xu and Lee, 2003; Gardina *et al.*, 2006; He *et al.*, 2007; Xi *et al.*, 2008).

Several years ago, Affymetrix, Inc. introduced an innovative array platform called 'GeneChip Exon 1.0 ST Arrays', which is commonly known as 'exon arrays'. Unlike a conventional microarray platform, which is designed to target the 3′-end of mRNAs (e.g. HG-U133 Plus2 arrays), exon arrays can measure the expression of each exon. With more than 5.5 million features on each array, more than 90% of the exons are covered with four probes to comprehensively cover the annotated genomes of human, rat and mouse (http://media.affymetrix.com/support/technical/datasheets/exon_arraydesign_datasheet.pdf). To provide such comprehensive coverage, probes are designed from empirically supported and predicted sequencing using databases (e.g. Ensembl) and prediction programs (GENSCAN; http://genes.mit.edu/GENSCAN.html). With such design, it is possible to analyze the following subsets of probe sets:

- Core: RefSeq transcripts and full-length mRNAs
- Extended: Core + cDNA-based annotations
- Full: Extended + *ab-initio* gene predictions

For more information about the annotations, please refer to: http://www.affymetrix.com/support/technical/whitepapers/exon_probeset_trans_clust_whitepaper.pdf. As probe sets of the extended and full annotations often result in poor hybridizations, it is not recommended to analyze at the gene level with these annotations (http://www.affymetrix.com/support/technical/whitepapers/exon_gene_signal_estimate_whitepaper.pdf).

Compared with conventional 3′ microarrays, which are only usable for detecting the expression level of entire genes, analyzing exon arrays is more difficult and requires the manual inspection of each potential candidate. To overcome this problem, we built an easy-to-use web interface called 'Exon Array Analyzer (EAA)' (http://EAA.mpi-bn.mpg.de/) (Gellert *et al.*, 2009). EAA is an all-in-one tool to analyze exon arrays, which includes the preprocessing of raw CEL files, the detection of alternative spliced exons, the filtering of false positive results, a genome viewer and sequence information for primer design. All the above features are available to the user without any programming skills. Because EAA runs completely on a web browser, no installation is required. For the purpose of finding tissue-enriched genes and tissue-specific exons, exon arrays of 11 different murine tissues were downloaded from Affymetrix (http://www.affymetrix.com). Using our EAA, the datasets were normalized and compared among each other. Furthermore, mouse exon array data from the BioGPS project (Wu *et al.*, 2009) (http://www.ncbi.nlm.nih.gov/projects/geo/query/acc.cgi?acc=GSE15998) were downloaded and analyzed by our EAA to add more datasets.

4.3.4 Mouse Atlas of Gene Expression

Serial analysis of gene expression (SAGE) was developed in the early 1990s to detect and quantify mRNA in a sample of interest by utilizing small tags that correspond to fragments of those transcripts (Velculescu *et al.*, 1995). Unlike microarrays, SAGE does not require pre-existing sequences for probe construction, which allows *de novo* identification of novel genes. After sequencing of all tags, these sequences are mapped back to the mRNA sequence databases [e.g. SAGEmap (http://www.ncbi.nlm.nih.gov/projects/SAGE/), SAGE Genie (http://cgap.nci.nih.gov/SAGE) (Boon *et al.*, 2002)] to identify the corresponding genes. Simultaneously, the frequency of tags is determined, allowing the quantification of expression levels in comparison with a control sample (Wang, 2007). Sometimes it is difficult to identify tags due to their short length. Therefore, since its introduction in 1995, the length of small fragments of sequences has been increased from 10 (short SAGE) to 17 (long SAGE), 21 [Robust-LongSAGE tag (Gowda *et al.*, 2004)] and 26 [Super-SAGE tag (Matsumura *et al.*, 2005)].

The Mouse Atlas of Gene Expression project (www.mouseatlas.org) generated 202 SAGE libraries from 198 tissues and different developmental stages to provide a comprehensive catalog of genes expressed throughout the development of various organs of mouse strain C57BL/6J, the most commonly used strain in labs (Siddiqui *et al.*, 2005). Information about all the libraries can be found at: http://www.mouseatlas.org/data/mouse/. This is by far the most comprehensive collection of SAGE expression profiles currently available. For example, in the case of heart, ten libraries from five developmental stages [whole hearts from embryonic day (E)8.5 embryos; atria, atrioventricular canals, bulbus cordis and left ventricles from E9.5 embryos; atria, atrioventricular canals and ventricles from E11.5 embryos; atria and ventricles from E12.5 embryos] are available. In each library, experiments were performed with the 10-bp SAGE or the more reliable 17-bp SAGE method.

To obtain genes expressed during the development of a tissue, all tags from the corresponding libraries of the target tissue were pooled and annotated with 'SAGEmap_Mm_NlaIII_10_best' or 'SAGEmap_Mm_NlaIII_17_best' from NCBI's SAGEmap (http://www.ncbi.nlm.nih.gov/projects/SAGE/) (both libraries constructed from UniGene Build 170) (Lash *et al.*, 2000). All tags with count > 1 were accepted as genes expressed during the development of the target tissue.

4.3.5 Architecture of C-It

The above mentioned transcriptomics datasets were merged to screen for evoluationarily conserved, tissue-enriched genes, which are uncharacterized based on the number of publications (PubMed) for the target tissue by utilizing Medical Subject Headings (MeSH) terms. To find the number of publications related to each gene, Entrez Gene's 'gene2pubmed' dataset was utilized. First, the number of Entrez's GeneIDs (GIs) that are listed under each PubMed ID (PMID) was counted and classified. Publications that include more than 100 GIs were considered as articles that report large screening results (e.g. microarrays, next-generation sequencing) and excluded from the count for the number of publications for each gene. Next, PMIDs were categorized according to the MeSH classification (http://www.nlm.nih.gov/mesh/). As briefly described in section 2.2.7, MeSH provides a comprehensive collection of a controlled vocabulary thesaurus for the purpose of indexing journal articles and books to facilitate a search based on terms used in Medical Science. It is organized in a hierarchical tree structure to permit a search at various levels of specificity (http://www.nlm.nih.gov/mesh/trees.html). Because every journal article in MEDLINE and PubMed is indexed with headings and subheadings for searching purposes, such headings are also linked to MeSH terms. Using this connection, the PMIDs identified above and the MeSH term for a particular tissue were cross matched to the number of publications for each gene. If this number was less than 2, the GI (gene) was defined as being 'uncharacterized'.

Figure 4.1 shows how each dataset has been connected to each other to build 'C-It' (http://C-It.mpi-bn.mpg.de) (Gellert *et al.*, 2010).

C-It was built using the relational database MySQL (http://www.mysql.com/). Detailed information about MySQL and the specific usage goes well beyond the scope of this book. Readers are advised to consult the many excellent introductory books and articles both on and offline about the database architecture as well as MySQL. The reason behind using MySQL for C-It is that it is free and expandable so that other datasets (e.g. proteomics, metabolomics) can be added later. We are currently extending C-It by incorporating datasets of stem cells and non-coding RNAs (ncRNAs) (Uchida *et al.*, 2011).

4.3.6 Example: evolutionarily conserved, heart-enriched, uncharacterized genes

Let us now use C-It to screen for evolutionarily conserved, heart-enriched, uncharacterized genes (Figure 4.2). Press the [Submit Query] button to proceed to the next page (Figure 4.3).

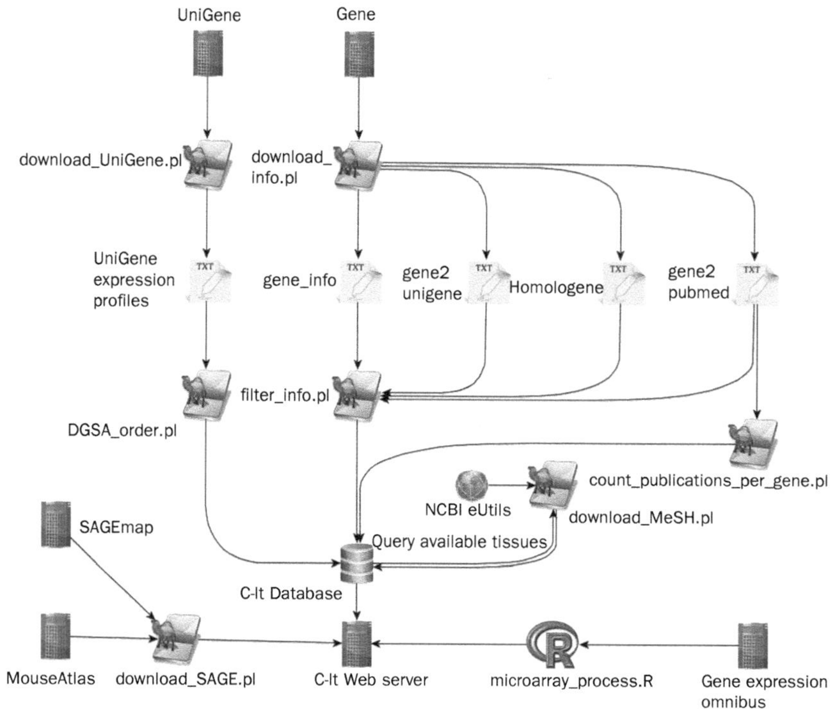

Figure 4.1 The database scheme for C-It

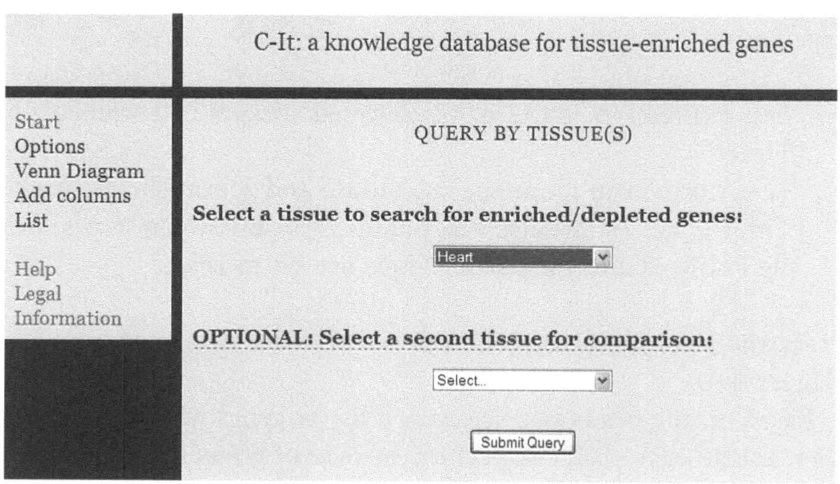

Figure 4.2 Start page of C-It. 'Heart' is selected as an example

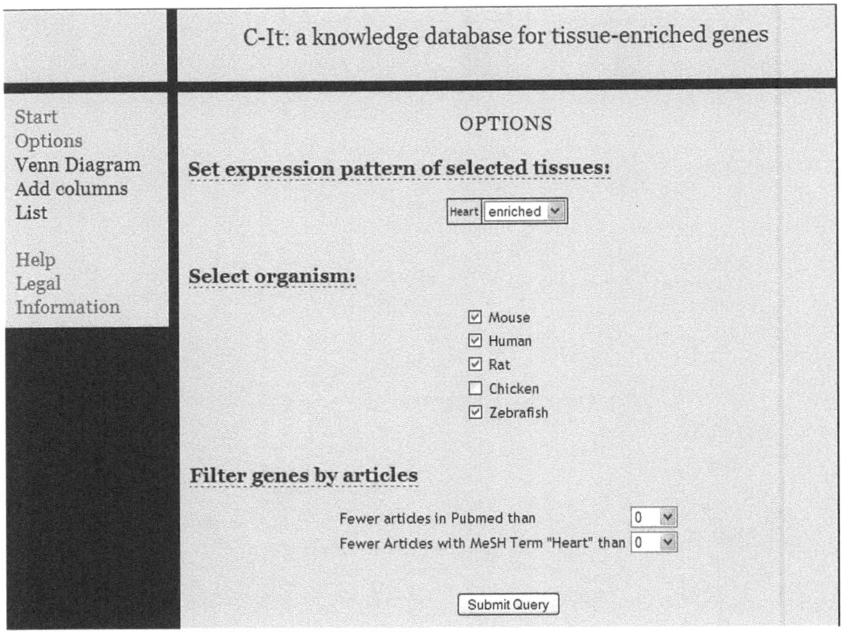

Figure 4.3 Selection page for options. Here, a user can specify filters

In the new page, you can select the conditions in which to filter for the genes of interest. In the above example, the following are specified:

- 'Set expression pattern of selected tissues': 'enriched' for 'Heart'
- 'Select organism': human, rat, mouse and zebrafish are selected
- 'Filter genes by articles': to obtain truly unknown genes, both the PubMed and the MeSH terms are set to zero.

Press the [Submit Query] button to proceed to screen for genes (Figure 4.4).

Based on the filters you selected, a list of genes will be generated. To make an easy visual inspection, instead of going through a list of genes that can be very long, a Venn diagram is drawn. As can be seen from Figure 4.3, there is only one gene [*Klhl38* (kelch-like 38)

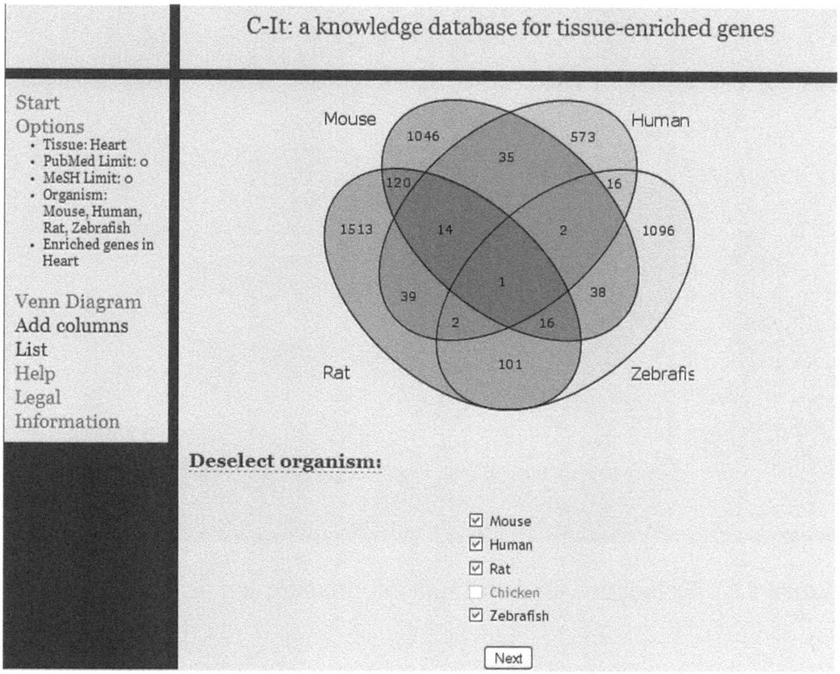

Figure 4.4 Venn diagram of overlapping genes

(*Drosophila*)] that is conserved among the four species. In this case, to increase the number of selected genes, you can 'deselect' organisms. For example, omit 'rat' and press the [Next] button (Figure 4.5). On this page, additional information to be displayed can be selected. For the purpose of this example, all the columns will be selected. Press the [Next] button to continue (Figure 4.6).

The result of the screening is shown in Figure 4.6. Because the results table can be very large, scroll bars are provided. Furthermore, a column can be sorted, if applicable, to make the viewing of results easier. Hyperlinks are provided to display further information in a separate window. For example, if you click on '1.81', which is the score in the 'SymAtlas z-value in Heart' for the first gene 'storkhead box 2 (*Stox2*)', the expression profiles of this gene are displayed as shown in Figure 4.7. From this window, the expression profiles of

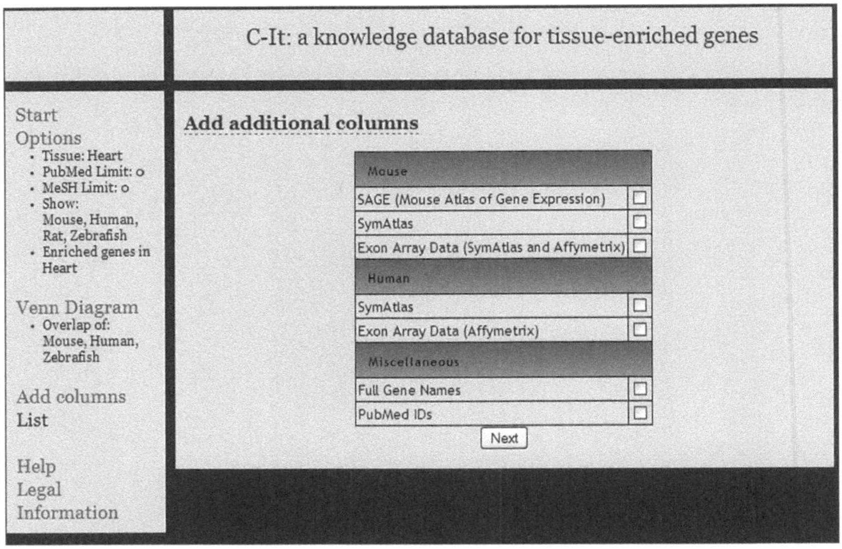

Figure 4.5 Selections for additional information

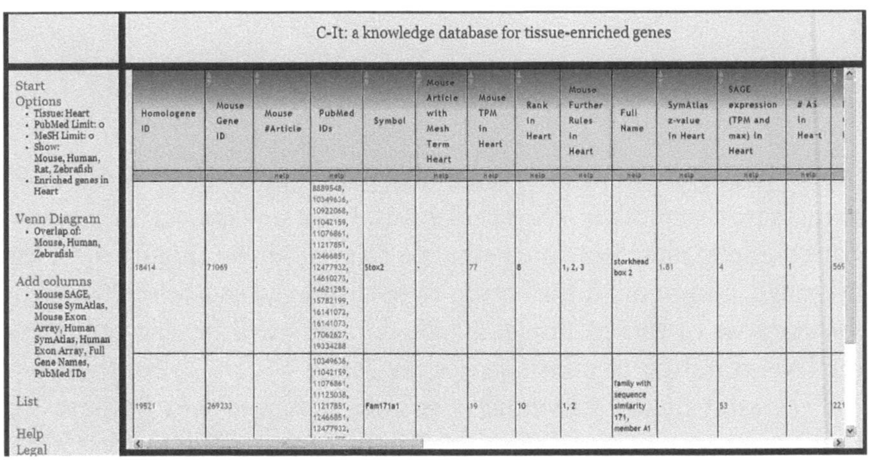

Figure 4.6 Results table

homologous genes can be viewed by clicking on the tabs with the organism's name.

C-It includes many datasets and is linked to various publicly available databases to allow the user to search for evolutionarily conserved, tissue-enriched, uncharacterized genes.

C-It

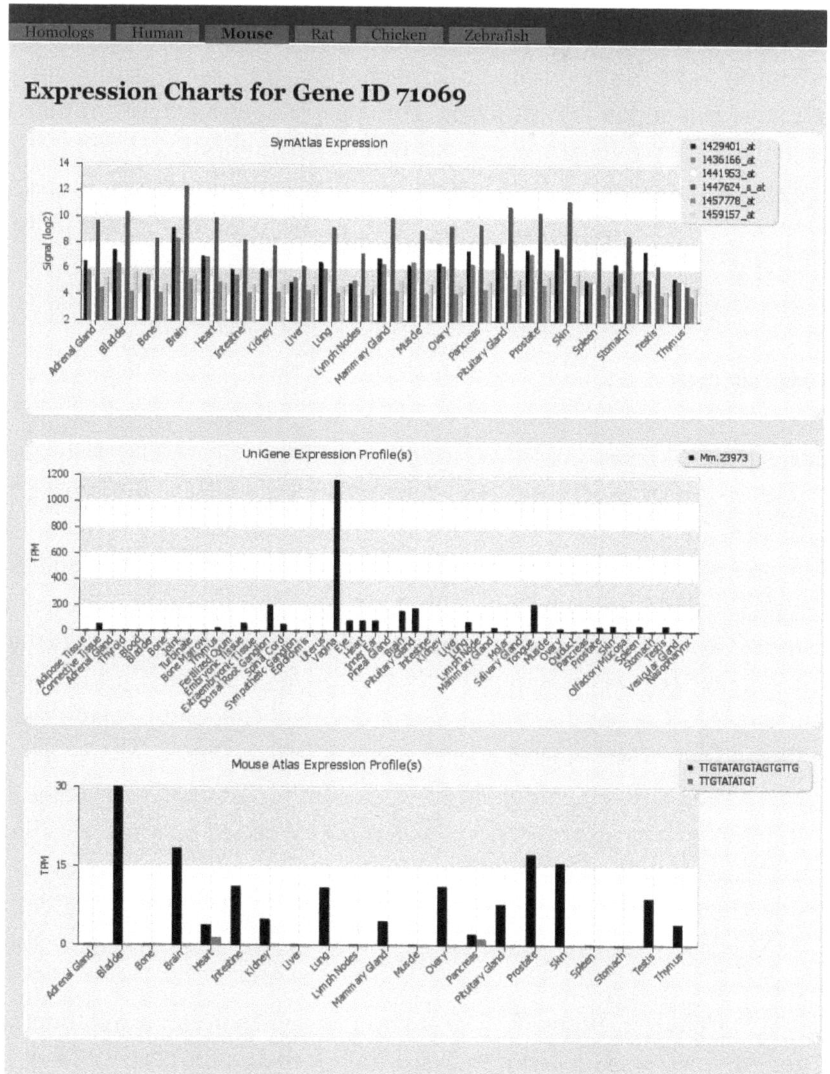

http://ci.mpi-bn.mpg.de/detail_index.php?tissue=heart&id=exp_mouse&homolog_id=18414&single_gene=71069[8/16/2011 1:00:40 PM]

Figure 4.7 Expression profiles of *Stox2*

4.4 Conclusion

In this chapter, the concept of *in silico* screening to search for evolutionarily conserved, tissue-enriched, uncharacterized genes was described, and our C-It knowledge database was used as an example. The most important thing to bear in mind is that the goal of *in silico* screening is to data-mine the information (i.e. biological data) to extract knowledge by utilizing the past experience of numerous biologists. As we have already shown experimentally (Uchida *et al.*, 2009), our *in silico* screening can be utilized to find novel genes. In the next chapter, the methods to validate such genes experimentally will be described.

References

Bitton DA, Okoniewski MJ, Connolly Y, Miller CJ. (2008) Exon level integration of proteomics and microarray data. *BMC Bioinformatics* 9: 118.

Blencowe BJ. (2006) Alternative splicing: new insights from global analyses. *Cell* 126(1): 37–47.

Boon K, Osorio EC, Greenhut SF, *et al.* (2002) An anatomy of normal and malignant gene expression. *Proc Natl Acad Sci USA* 99(17): 11287–92.

Faustino NA, Cooper TA. (2003) Pre-mRNA splicing and human disease. *Genes Dev* 17(4): 419–37.

Gardina PJ, Clark TA, Shimada B, *et al.* (2006) Alternative splicing and differential gene expression in colon cancer detected by a whole genome exon array. *BMC Genomics* 7: 325.

Gautier L, Cope L, Bolstad BM, Irizarry RA. (2004) affy – analysis of Affymetrix GeneChip data at the probe level. *Bioinformatics* 20(3): 307–15.

Gellert P, Uchida S, Braun T. (2009) Exon Array Analyzer: a web interface for Affymetrix exon array analysis. *Bioinformatics* 25(24): 3323–4.

Gellert P, Jenniches K, Braun T, Uchida S. (2010) C-It: a knowledge database for tissue-enriched genes. *Bioinformatics* 26(18): 2328–33.

Gowda M, Jantasuriyarat C, Dean RA, Wang G. (2004) Robust-LongSAGE (RL-SAGE): a substantially improved LongSAGE method for gene discovery and transcriptome analysis. *Plant Physiol* 134(3): 890–7.

He C, Zuo Z, Chen H, *et al.* (2007) Genome-wide detection of testis- and testicular cancer-specific alternative splicing. *Carcinogenesis* 28(12): 2484–90.

Hutten EH. (1960) Physics and biology. *Br J Philos Sci* 11(42): 101–8.

Johnson JM, Castle J, Garrett-Engele P, *et al.* (2003) Genome-wide survey of human alternative pre-mRNA splicing with exon junction microarrays. *Science* 302(5653): 2141–4.

Lash AE, Tolstoshev CM, Wagner L, *et al.* (2000) SAGEmap: a public gene expression resource. *Genome Res* 10(7): 1051–60.

MAQC Consortium, Shi L, Reid LH, Jones WD, *et al.* (2006) The MicroArray Quality Control (MAQC) project shows inter- and intraplatform reproducibility of gene expression measurements. *Nat Biotechnol* 24(9): 1151–61.

Matsumura H, Ito A, Saitoh H, *et al.* (2005) SuperSAGE. *Cell Microbiol* 7(1): 11–18.

Modrek B, Resch A, Grasso C, Lee C. (2001) Genome-wide detection of alternative splicing in expressed sequences of human genes. *Nucleic Acids Res* 29(13): 2850–9.

Pan Q, Shai O, Lee LJ, Frey BJ, Blencowe BJ. (2008) Deep surveying of alternative splicing complexity in the human transcriptome by high-throughput sequencing. *Nat Genet* 40(12): 1413–15.

Quackenbush J. (2002) Microarray data normalization and transformation. *Nat Genet* 32 Suppl.: 496–501.

Siddiqui AS, Khattra J, Delaney AD, *et al.* (2005) A mouse atlas of gene expression: large-scale digital gene-expression profiles from precisely defined developing C57BL/6J mouse tissues and cells. *Proc Natl Acad Sci USA* 102(51): 18485–90.

Su AI, Cooke MP, Ching KA, *et al.* (2002) Large-scale analysis of the human and mouse transcriptomes. *Proc Natl Acad Sci USA* 99(7): 4465–70.

Sumanas S, Larson JD. (2002) Morpholino phosphorodiamidate oligonucleotides in zebrafish: a recipe for functional genomics? *Brief Funct Genomic Proteomic* 1(3): 239–56.

Uchida S, Nishida Y, Satou K, Muta S, Tashiro K, Kuhara S. (2005) Detection and normalization of biases present in spotted cDNA microarray data: a composite method addressing dye, intensity-dependent, spatially-dependent, and print-order biases. *DNA Res* 12(1): 1–7.

Uchida S, Schneider A, Wiesnet M, *et al.* (2009) An integrated approach for the systematic identification and characterization of heart-enriched genes with unknown functions. *BMC Genomics* 10: 100.

Uchida S, Gellert P, Braun T. (2011) Deeply dissecting stemness: making sense to non-coding RNAs in stem cells. *Stem Cell Rev* [Epub ahead of print]. Doi: 10.1007/s12015-011-9294-y. http://www.springerlink.com/content/u42387202r4743k7/

Velculescu VE, Zhang L, Vogelstein B, Kinzler KW. (1995) Serial analysis of gene expression. *Science* 270(5235): 484–7.

Wang ET, Sandberg R, Luo S, *et al.* (2008) Alternative isoform regulation in human tissue transcriptomes. *Nature* 456(7221): 470–6.

Wang GS, Cooper TA. (2007) Splicing in disease: disruption of the splicing code and the decoding machinery. *Nat Rev Genet* 8(10): 749–61.

Wang SM. (2007) Understanding SAGE data. *Trends Genet* 23(1): 42–50.

Wu C, Orozco C, Boyer J, *et al.* (2009) BioGPS: an extensible and customizable portal for querying and organizing gene annotation resources. *Genome Biol* 10(11): R130.

Xi L, Feber A, Gupta V, *et al.* (2008) Whole genome exon arrays identify differential expression of alternatively spliced, cancer-related genes in lung cancer. *Nucleic Acids Res* 36(20): 6535–47.

Xu Q, Lee C. (2003) Discovery of novel splice forms and functional analysis of cancer-specific alternative splicing in human expressed sequences. *Nucleic Acids Res* 31(19): 5635–43.

Xu Q, Modrek B, Lee C. (2002) Genome-wide detection of tissue-specific alternative splicing in the human transcriptome. *Nucleic Acids Res* 30(17): 3754–66.

Yeo G, Holste D, Kreiman G, Burge CB. (2004) Variation in alternative splicing across human tissues. *Genome Biol* 5(10): R74.

5

Experimental validation: concepts and practice

Abstract: *In silico* screening allows you to make a list of interesting predictions, and these need to be tested experimentally. In this chapter, I describe the methods you could utilize to validate your predictions – a 'proof-of-principle (or concept)' approach. As the type of experiments to be performed varies case-by-case, a universal set of experiments are described, with reference to more specialized experimental methods.

Keywords: expression, experimental validation, proteomics, transcriptomics

5.1 Introduction

As described in Chapter 1, about 27% of human genes remain undescribed (Table 1.1). This proportion increases significantly in model organisms, such as rat (49%) and chicken (74%). The functions of many genes are therefore still unknown. These genes are collectively called 'function-unknown genes' or 'novel genes'. When we examine their expression patterns, as shown in the previous chapter, some genes are found to be expressed exclusively or highly in a particular tissue (e.g. heart) compared with others. It seems unlikely that any organism would spend an enormous amount of energy (e.g. ATP) generating unwanted proteins from such genes, here termed 'tissue-enriched, function-unknown genes'. Furthermore, some of such genes are well conserved across species, indicating that they must have important functions related to the survival and/or tissue homeostasis of an organism. An obvious conclusion that we could draw by looking at such a list of evolutionarily conserved, tissue-enriched, function-unknown genes is that functional, biological studies targeting such genes are lacking. Thus, additional biological experiments are necessary to elucidate the functions of such genes.

In this chapter, I describe a series of biological experiments that one could use to elucidate the functions of a gene [or non-coding RNA (ncRNA)] that is evolutionarily conserved, tissue-enriched and function-unknown. I describe the concept behind each method and some tips for performing such experiments. I will only briefly touch on some of the routine experimental methods, so this chapter is by no means a comprehensive guide to Experimental Biology. Useful protocols are available in the scientific literature [e.g. Cold Spring Harbor Protocols (http://cshprotocols.cshlp.org/), Nature Methods (http://www.nature.com/nmeth/index.html), Nature Protocols (http://www.nature.com/nprot/index.html) to name but a few] and via databases, such as Protocol Online (http://www.protocol-online.org/). The *Journal of Visualized Experiments* (http://www.jove.com/) is unique in providing a video to follow each step of the experiment. Of course, asking your colleagues will be the easiest way to acquire the necessary protocols and skills to answer your research questions, and corresponding authors can provide details of any protocol used in a particular experiment. Every researcher will be happy to know that his or her article has been read by others and is of interest. Readers are advised to critically access the necessity of performing experiments listed in this chapter to answer the question you are interested in.

5.2 Proposed experimental strategy

Although there are plenty of evolutionarily conserved, tissue-enriched, function-unknown genes, we cannot simply pick one randomly and study its function; a biologist could spend all of his or her lifetime studying just one gene. Given that there are close to

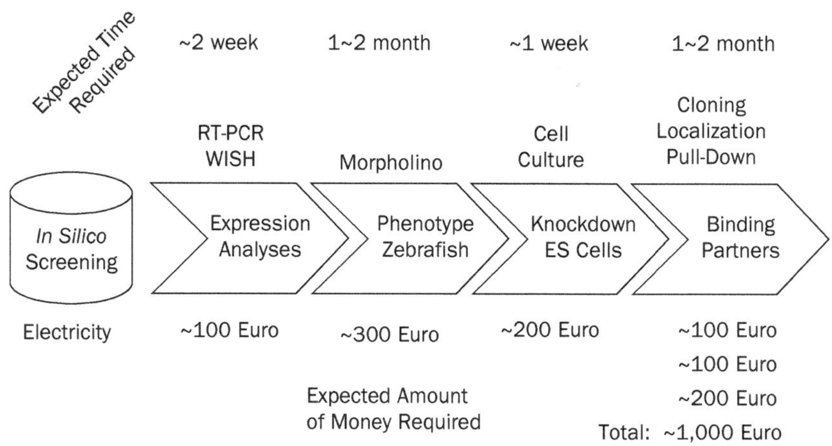

Figure 5.1 Flow chart of the 1,000 Genes Project. The expected time and amount of money required are estimates. Depending on the conditions you are working with, the expected values will change accordingly. The above series of experiments will be followed by working with knockout mice in our case

1,000 evolutionarily conserved, heart-enriched, uncharacterized (function not known in the tissue of interest, 'heart' in our case) genes (Uchida *et al.*, 2009) (thus, the name of our '1,000 Genes Project'), this is equivalent to the life journeys of a thousand biologists! We simply do not live long enough to see the end of such efforts. Therefore, a more systematic approach must be employed to elucidate the functions of such genes. Figure 5.1 shows the strategy that we are currently employing in my group.

One obvious question is which gene to focus on. If you ask this question to your boss or PhD advisor, the answer might well be one that can be published in a high-impact factor journal. But life is not so easy. If it is so easy to find such a gene and elucidate its function, then clearly it would not have been considered to be 'function unknown' based on its publication record, and this is often forgotten by researchers. The following is a set of guidelines to be considered when selecting a gene for further study:

- *Which particular cell type are you interested in?*
  - Do not assume that an organ (e.g. heart) is made up of one type of cell (e.g. cardiomyocytes). All organs contain some types of endothelial cells (not all of which are the same) and fibroblasts (again, not all of them are the same). Rinn *et al.* (2006) discuss fibroblasts from different parts of the human body.
  - Simply due to the lack of the appropriate marker for a certain cell type (e.g. fibroblasts; Neilson, 2010), few datasets are currently available with cell-type-specific information. The NIH-funded project called 'Genotype-Tissue Expression (GTEx)' (https://commonfund.nih.gov/GTEx/) will add more information concerning this aspect in the near future.
  - Given the above two points, you should consider looking for an organ of interest bearing in mind that depending on the cell type in which the gene of interest is expressed, you might see similar phenotypes in organs other than your target when this gene's activity is manipulated (e.g. knockout, knockdown or over-expression). You might get an idea about the expression pattern of the gene of interest for a particular cell type by accessing the Human Protein Atlas database (http://www.proteinatlas.org//index.php).
- *Which particular cellular process (e.g. apoptosis pathway) are you interested in?*
  - You can simply filter for those genes whose binding partners are involved in such processes using published results of yeast-two-hybrid or mass spectrometry-based protein–protein interaction datasets, which are described in section 2.3.2.4.4.
  - By accessing the Gene Expression Omnibus (GEO) or ArrayExpress, you can find relevant microarray studies for such processes and screen for the gene of interest to see if its expression changes significantly under the cellular process of interest.
- *Are there any functional domains of the protein coded by the gene of interest?*

- Some functional domains are well known for a certain cellular process. Therefore, you might gain an idea about the function of the gene of interest by examining the information about proteins with similar functional domains.
- Such functional domain information can be found through Pfam or InterPro, which are described in section 2.3.2.4.2.
- By assuming that such a functional domain is conserved across species, you may find relevant information in lower organisms, such as yeast or *E. coli*.

- *How evolutionarily conserved is the gene of interest?*
  - By accessing to the databases listed in section 2.3.2.1, you can obtain a list of homologous genes. If the gene of interest is conserved from very primitive species (e.g. *E. coli*), it is likely that the function of this gene is important for the survival of an organism.
  - Conceptually, by going down the phylogenetic tree based on levels of gene sequence similarity (as in the case of homologues) and anatomy (e.g. the contractile activity of cardiomyocytes evolved from those of skeletal and smooth muscles), you could utilize lower species experimentally. Usually, for larger model organisms, the more money you need to perform an experiment. For example, consider using pigs versus mice for cardiovascular research. The advantage of using pig hearts over murine hearts is that the former is much closer to the human heart. However, it costs far more to feed and take care of (e.g. space for housing) pigs than mice.

As you can easily imagine, the above guidelines can be followed *in silico* using the databases list in Chapter 2 and processing the information contained in such databases using regular expression within Perl as described in Chapter 3. Alternatively, you could use our C-It knowledge database (http://C-It.mpi-bn.mpg.de) (Gellert *et al.*, 2010) to search for evolutionarily conserved (zebrafish, chicken, mouse, rat and human), tissue-enriched, uncharacterized genes (Chapter 4).

5.3 Step 1: validation of expression patterns

After you have successfully screened for an interesting set of genes *in silico*, next you need to validate your predictions. In Biology, many researchers use terms such as 'proof-of-principle' or 'proof-of-concept' to describe this type of study. In reality, no biologist will trust such *in silico* screenings and predictions, unless they can be validated experimentally. Such experiments are clearly far more complex than simply turning on a computer and are time consuming. Furthermore, many of the experiments described below must be done in accordance with the laws and regulations of your local and national authorities. Thus, the first thing you need to do is to check with your advisor, boss, school, department, institute, etc., to make sure that you are allowed to perform the experiments.

5.3.1 At the RNA level

As the above described *in silico* screening is based largely on transcriptomics datasets, the first step is to validate mRNA expression levels. Compared with proteomics, validation via the transcriptomics approach is much easier to perform. In the following subsections, transcriptomics methods are described in detail.

5.3.1.1 RT-PCR

Having screened (*in silico*) for a gene of interest, you now need to confirm that this gene is expressed in the tissue and/or cell type of interest. This can be easily done by the reverse-transcriptase polymerase chain reaction (RT-PCR). Alternatively, you could perform Northern blotting, but such experiments require far more RNA than PCR-based experiments. The materials to be used vary depending on your particular research goal, but you should consider preparing different types of samples for comparison. In my group, when we screen for tissue-enriched genes, we use RNA from 15 tissue types (brain, eyes, lungs, heart, liver, kidneys, spleen, stomach, small intestine, large intestine, white fat, brown fat, skeletal muscle, testes and uterus – see Uchida *et al.*, 2009) from adult mice, which are isolated using TRIzol® reagent [Invitrogen (Life Technologies), Catalog No. #15596] and synthesis of first-strand cDNA using SuperScript™ Reverse Transcriptase [Invitrogen (Life Technologies), Catalog No. #18064] by following the protocol specified by the manufacturer. Regarding homogenization, although some researchers prefer and recommend using blender-type homogenizers (e.g. Polytron), in the case of very limited amount of a sample (e.g. biopsy sample, which is only the size of a tip of a needle), simply pipetting it in TRIzol reagent works best without significant loss of material. An informative website regarding cell and tissue dissociations is provided by Worthington Biochemical Corporation (http://www.worthington-biochem.com/tissuedissociation/default.html), which you might want to refer to before proceeding with your experiments.

Unlike many other laboratories, we do not treat RNAs with DNase I to cleave genomic DNA. The TRIzol reagent manual states: 'For use in the polymerase chain reaction (PCR*), treatment of the isolated RNA with amplification grade DNase I (Cat. No. 18068) is recommended when the two primers lie within a single exon.' In other words, as long as a primer pair for RT-PCR is designed to span two exons, the DNase treatment is not required. The advantages of avoiding DNase treatment are fewer steps and less costs as well as less RNA damage. Avoiding DNase treatment can be very beneficial both experimentally and economically.

We use the following steps in primer design:

1. Obtain the sequence information, including exons, from the Ensembl database.
2. Copy and paste the sequence to a NotePad or another text editor. Alternatively, you could utilize an application designed for sequence analysis (e.g. Lasergene's SeqBuilder software) so that you can mark the exon boundaries. We usually mark all even numbered exons (e.g. Exons 2, 4, 6, ..) with a red box.
3. Pick about 20 bases in one exon or between an exon junction, which has 10–12 GC content. It is advisable that the primer starts with guanine (G) or cytosine (C) and to avoid having more than three repetitions of G and/or C.

4. Blast (http://blast.ncbi.nlm.nih.gov/Blast.cgi) the sequence to determine specificity.
5. Feed all of the information (full mRNA sequence and the sequence from Step 3) to Primer3 (http://frodo.wi.mit.edu/), which is by far the most useful primer design web interface. The parameter settings are as follows:
   a. 'Product Size Ranges': 100–200

   The following are the reasons for designing the PCR product to be between 100 and 200 base pairs (bp):

   - In some cases, you might note that a primer dimmers (<40 bp). To avoid an overlap of size, it is better to keep the PCR products above 50 bp. Also, if the product is less than 100 bp, it is often more expensive to find an appropriate molecular marker (DNA ladder) that can specify the size of interest to see if the PCR experiment worked when run on an agarose gel.
   - It depends on the type of DNA polymerase, but a rule of thumb for *Taq* polymerase is extension of 1000 bases (1 kb) per minute. By specifying the PCR product to be less than 200 bp, the size recommended for most real-time RT-PCR machines, you could perform the PCR reaction in less time (200 bp ÷ 1 kb/min = 12 s extension time). If you do not have so much time for the PCR reaction to finish, the following reaction scheme should work with the above product size (we use the following PCR scheme reasonably often): 95 °C for 2 min, 94 °C for 10 s (denaturing), 60 °C* for 10 s (annealing), 72 °C for 20 s (extension) and 72 °C for 7 min (*the annealing temperature depends on your primer pair).

   In most PCR machines, the 40-cycle reaction with the above scheme should take less than 1 h. Sometimes, an experiment can be performed in 40 min or less depending on the number of cycles and the performance of the PCR machine.

   b. 'Number To Return': 500

c. Paste the primer sequence that you designed in Step 2 to either 'Pick left primer, or use left primer below:' for the forward primer or 'Pick right primer, or use right primer below (5′ to 3′ on opposite strand):' for the reverse primer

d. Adjust 'Primer Tm' accordingly (I usually design the annealing temperature of a primer pair to be around 60 °C).

6. Blast the sequence to determine specificity. By comparing the Blast results of the forward and reverse primers, there should be only one match; that is, the gene of interest. There may be isoforms present for the gene of interest that result in multiple overlaps between the two Blast results. As long as there is no overlap of results for genes other than your target, the primer pair that you designed is ready to be ordered for use.

Note that the above procedure works only for a gene with more than one exon. Alternatively, instead of designing one primer spanning two exons, you can simply feed in the whole mRNA sequence to Primer3 to let it design an appropriate primer pair for you, which is simpler to work with.

As long as the organism that you are working with is well characterized (by means of genome sequencing), you will obtain the right PCR product, given that you have optimized the annealing temperature of the PCR reaction. However, if you are in doubt, you could clone the product into a vector. For this purpose, we prefer to use the pGEM®-T Easy Vector Systems (Promega, Catalog No. A1360), which has a 3′-terminal thymidine (T) to both ends of the insertion site. Based on the fact that *Taq* DNA polymerase adds a 3′-terminal deoxyadenosine (A) to a PCR product, TA-cloning can be performed. For details about this vector, please refer to the manual provided by Promega (http://www.promega.com/resources/protocols/technical-manuals/0/pgem-t-and-pgem-t-easy-vector-systems-protocol/). This procedure is tedious and time consuming. Alternatively, you could perform restriction enzyme digestions to check the expected sizes. For this, the following procedure works efficiently in my group:

1. Find unique restriction enzyme sites (i.e. the site that is cut only once) from the expected PCR product. This can be done easily by using a feature installed in the software product designed for DNA sequence analysis (e.g. Lasergene's SeqBuilder software).
2. Run a PCR reaction in a 20-μL volume for 40 cycles to generate a sufficient amount of the product.
3. Run 5 μL of the PCR reaction on a 2% agarose gel. Make sure that there is only one band.
4. Pick two restriction enzymes, which each enzyme cuts once in the product region, to see if the PCR product amplified is what you expected. In my group, the following restriction enzymes work very well: *Alu*I, *Bgl*II, *Bpm*I, *Bsa*JI, *Eco*RI, *Eco*RV, *Msp*I, *Nla*III, *Nla*IV, *Pvu*II, *Sal*I and *Xba*I. Perform the following digestions (BSA, bovine serum albumin):

	Without BSA	With BSA
PCR product	5 μL	5 μL
10 × Buffer	1	1
BSA	0	1
Water	3.5	2.5
Restriction enzyme	0.5	0.5
Total	10 μL	10 μL

5. Run all the reaction, together with the remaining ~5 μL of PCR reaction (you ran the PCR reaction in a 20-μL volume, in which 5 μL was used in Step 3 and 10 μL in Step 4), on a 2% agarose gel to see the shift in sizes.

If you would like to perform a more quantitative analysis of the gene of interest, you should perform real-time RT-PCR [also known as quantitative RT-PCR (qRT-PCR)]. For details about this technology, please refer to the excellent review by VanGuilder *et al.* (2008). As indicated in the above steps, as long as the size of a PCR product is less than 200–300 bp, the same primer pair as designed above can be used. The reason for the shorter product size is based on the characteristics of the commonly used dye SYBR Green

(Morrison *et al.*, 1998), which binds to double-stranded DNA but not to single-stranded DNA; thus, it is called an 'intercalator'. When it is bound to the double-stranded DNA, it generates very bright fluorescence (much more brightly than ethidium bromide does). The longer the product size, the more signals will be emitted. Thus, it is harder to quantify (Bustin, 2000). For this reason, I prefer to use the following scheme for normalization using a standard curve to analyze at the absolute level, instead of using the $2^{-\Delta\Delta CT}$ method at the relative level (VanGuilder *et al.*, 2008):

1. Using a primer pair of your choice, perform a test run (only one reaction) for the real-time RT-PCR.
2. Run all the volume on a 2% agarose gel and cut out the band.
3. Purify the band using a conventional DNA extraction kit [e.g. NucleoSpin® Extract II (Macherey-Nagel, Catalog No. 740609)].
4. Quantify the DNA (it should be somewhere between 1 and 10 ng/µL) using a NanoDrop (http://www.nanodrop.com/) or similar machine.
5. Make serial dilutions as follows for a standard curve: (A) 1 ng/µL, (B) 100 pg/µL, (C) 10 pg/µL, (D) 1 pg/µL, (E) 100 fg/µL, (F) 10 fg/µL, (G) 1 fg/µL, (H) 100 ag/µL (for settings of the real-time RT-PCR machine, I use 'femtogram' so this should be '0.1 fg/µL').
6. Use the above eight dilutions to draw a standard curve.

We perform the above procedure for all primer pairs, including an internal control. Regarding internal controls, readers are referred to the excellent article by Vandesompele *et al.* (2002). In my group, mouse *Actb* (β-actin) works very well – forward primer: CAACGAG CGGTTCCGATG, and reverse primer: GCCACAGGATTCCA TACCCA, which produce a 67-bp product annealing at 60 °C (Uchida *et al.*, 2007).

By performing (real-time) RT-PCR experiments, you should be able to validate the results of *in silico* screening experimentally. It might help you greatly to perform such an assay for a series of cell lines that you have so that you have an idea about which cell lines express your gene of interest. If a cell line expresses your gene of interest, then you could use this cell line to screen for binding partners of the protein that your gene of interest encodes as well as to check for the specificity of an antibody against your protein. If a cell line does not express your gene of interest, you could utilize it for gain-of-function assays because, in a sense, this cell line does not have your gene in the current status. By introducing your gene into the system, it might alter the phenotypes of the cells.

5.3.1.2 In situ hybridization

As I trained in the field of Developmental Biology, my interest lies in the developmental roles of evolutionarily conserved, tissue-enriched, uncharacterized genes. One of the first things that should be checked with regard to the development of an organism is the expression patterns of the gene of interest during development of the target organ. This can be done by *in situ* hybridization to detect mRNA of the gene of interest. When it is performed on the whole organism (i.e. embryos), this method is called whole-mount *in situ* hybridization (WISH). Because there are many excellent and detailed protocols available from various laboratories, I will not go into detail about this method; rather, I will list a few points which one should consider when performing this assay:

- Regarding a probe to detect mRNA of interest, you can use the same procedure as proposed above for RT-PCR, except changing the product size to be between 500 and 700 bp, which we have found works best.
- RNA is a stable molecule, but ribonuclease (RNase), which catalyzes the degradation of RNA, is around everywhere, including your saliva. Although diethylpyrocarbonate (DEPC)-treated solutions are used to avoid degradation of RNA, it is advisable to move quickly to protect RNA.
- Unfortunately, *in situ* hybridization is not a quantification method; thus, you must be careful about interpreting the results. The method indicates only whether the gene of interest is expressed at a certain point of development of an organism.

- The sensitivity of detection depends on the probe itself. Although the gene of interest may be expressed very highly in a target organ during development, it might take longer to detect its signal compared with other less well-expressed genes. In my group, as in the case of WISH, we sometimes keep murine embryos in the staining solution for several days in the refrigerator to develop enough color to separate the signal from background.
- In the case of WISH, it is always advisable to punch a hole in the brain cavity for embryos older than 9 days post-coitum to drain the staining solution trapped in the whole organism.

As listed in section 2.2.6, there are many databases with images of *in situ* hybridization experiments with which to compare your results. Although the target genes described here should be undescribed, you will be surprised to find so many images of function-unknown genes. This is because some of these datasets are created using high-throughput methods, such as automated probe constructions and robot-based washing. The purpose of such datasets is to provide information to a research community. A good example can be found through the Allen Brain Atlas (http://www.brain-map.org/), which used the RIKEN cDNA library (http://mouse.brain-map.org/documentation/index.html). Dr John Mattick, the pioneer in the field of long ncRNAs (lncRNAs), utilized such information to publish an elegant study to identify lncRNAs expressed in murine brain (Mercer *et al.*, 2008).

5.3.2 At the protein level

Although information about gene expression levels has led to important findings in the field of Functional Genomics, it is the proteins that are doing most of the jobs in tissue homeostasis and other important functions. Unlike RNA, proteins are rather unstable molecules. Not only do differences in temperature and pH cause the functionality of a protein to be altered, which may ultimately lead to its degradation, but there are many proteases present in the natural environment. This is clearly illustrated in the amount and number of protease inhibitors one must add to a reaction tube to prevent the target protein from degrading. Despite the above experimental obstacles, it is important to obtain information about the protein that your gene of interest encodes to confirm your *in silico* screening results.

5.3.2.1 Recombinant proteins

Regardless of the presence and availability of antibodies, you should always clone the gene of interest to make a recombinant protein for various reasons, including validation of the specificity of an antibody. In the case of multiple isoforms of your gene of interest, the information contained in public databases and viewers (e.g. UCSC Genome Browsers, Ensemble database) should be utilized to clone the longest isoform first and then to clone others once your *in silico* prediction has been validated. Below is the scheme that we use in my group to make recombinant proteins with green fluorescent protein (GFP) or V5/His tags:

1. Copy and paste the whole mRNA sequence of the longest isoform, if applicable, of the gene of interest to a sequence analysis program. Make sure to mark translation start (ATG) and stop (TAG, TAA or TGA) codons.
2. Check for the presence of restriction enzyme sites for *Bam*HI, *Sal*I and *Xho*I.
3. If the above mentioned restriction enzyme sites are not present in the full coding sequence, then pick 16 nucleotides from the translation start site and check for the melting temperature.
   - The free web interface provided by IIT Biotech (http://www.iit-biotech.de/iit-cgi/oligo-tm.pl) is very useful.
   - If the calculated melting temperature is somewhere between 58 and 64 °C, it is a good region for a primer.
   - If the temperature is too low, extend the selection of the nucleotides in the 3′ direction.
   - If the temperature is too high, decrease the number of nucleotides but not by less than 12.

4. Pick a reverse primer with or without [if a tag molecule (e.g. GFP) is attached at the C terminus of the recombinant protein] a stop codon with similar melting temperature.
5. Blast both forward and reverse primers to make sure that they amplify the target gene but not others.
6. Add the following sequences in front of the above designed primers:
   - Forward primer: AAA *GGATCC* GCCACC
     (*Bam*HI) Kozak sequence
   - Reverse primer: AAA *GTCGAC* or *CTCGAG*
     (*Sal*I) or (*Xho*I)
     Both *Sal*I and *Xho*I produce the compatible cohesive ends of each other.
7. Perform RT-PCR and clone into a TA-cloning vector (e.g. pGEM®-T Easy Vector Systems) to verify the fragment that you have amplified.
8. Cut out the fragment using the restriction enzymes that you designed.
9. Subclone into the following expression vectors:
   - **pEGFP-C1** (ClonTech, Catalog No. PT3028-5): adds enhanced GFP (eGFP) at the N terminus; cut with *Bgl*II (based on the compatible cohesive ends produced by *Bgl*II to those of *Bam*HI) and *Sal*I
   - **pEGFP-N3** (ClonTech, Catalog No. PT3054-5): adds eGFP at the C terminus; cut with *Bgl*II (based on the compatible cohesive ends produced by *Bgl*II to those of *Bam*HI) and *Sal*I
   - **pcDNA 3.1/V5-His A** [Life Technologies, Catalog No. V810-20]: adds V5 epitope and polyhistidine tag at the C terminus; cut with *Bam*HI and *Xho*I

With the above scheme, you can conveniently have three expression vector systems. The reason behind having GFP at both termini (pEGFP-C1 and pEGFP-N3) is to circumvent the usual problem that the GFP molecule (27 kDa) might interfere with the proper subcellular localization of a recombinant protein due to its large size (Palmer and Freeman, 2004). Furthermore, by having a His tag vector (pcDNA 3.1/V5-His), you can confirm whether the subcellular localizations of GFP-fused recombinant proteins are correct or not. The advantage of using GFP-fused proteins over Hig-tagged proteins is that live imaging of recombinant proteins can be made.

In some cases, it is now more convenient to order oligonucleotides to be synthesized than cloning the whole coding sequence by yourself using a conventional PCR method. Some companies, such as OriGene Technologies, Inc. (http://www.origene.com/), offer a wide collection of genes cloned into vectors with a variety of tagged proteins so that you can directly use them to transfect cells. Some of these companies use Gateway® recombination cloning technology [Life Technologies], which allows for shuttling of the insert DNA from one expression vector to another to simplify the cloning workflow. Because cloning takes time, by ordering from such companies, you can save time and effort. Given that you may have to perform several PCR experiments to obtain the gene of interest without any mutations, adding to the time you need to ligate to a vector and sequencing the insert, it might be cheaper and faster to buy one from such companies. Some principal investigators prefer such ordering over hiring and assigning a student or post-doc for such tasks, saving time and more money. Not only are private companies offering such deals. The RIKEN BioResource Center is distributing copies of full-length human cDNA clones, some of which are compatible with the Gateway® system (http://dna.brc.riken.jp/en/GNPcloneen.html). Therefore, if your gene of interest is of human origin, it will definitely save time and money to access such databases: the H-Invitational Database (H-InvDB; http://hinv.jp) and Human Gene and Protein Database (HGPD; http://www.HGPD.jp). You can obtain a full description and other initiatives from the Mammalian Gene Collection (MGC) website (http://mgc.nci.nih.gov/) as well as from the ORFeome Collaboration (http://www.orfeomecollaboration.org/) for model organisms.

After appropriate constructs for fusion proteins are made, the next step is to transfect the construct to cells to determine the subcellular localizations. There are a variety of methods available to transfect your constructs, including the classical calcium phosphate, electroporation, cationic liposome-mediated method [e.g. FuGENE® 6 (Roche Diagnostics) and Lipofectamine™ (Life Technologies) reagents] and the use of viral vectors. There are numerous articles that compare the efficiency and use of such methods (Cold Spring Harbor Protocol; 2006; doi:10.1101/pdb.prot4649), and I will not cover this aspect in detail here. The simplest way is to first try with a cell line that can be easily transfected [e.g. HEK 293 (http://www.atcc.org/ATCCAdvancedCatalogSearch/ProductDetails/tabid/452/Default.aspx?ATCCNum=CRL-1573&Template=cellBiology)], then move on to the cell lines that express your gene of interest to have all the necessary binding partners in the same cell. As we are interested in the development of heart, especially cardiomyocytes, we also work with chicken embryonic cardiomyocytes, which are easily electropolated and can contract rhythmically during cell division. Dr Joseph Sanger is the expert in using such a culture model to study how myofibril assembly occurs in vertebrate cross-striated muscles (called 'myofibril genesis'), and this is extensively reviewed in Sanger *et al.* (2005).

To determine subcellular localizations, there are various location-specific antibodies and dyes available. For example, Abcam has a very useful web page for this purpose: http://www.abcam.com/index.html?pageconfig=resource&rid=16. Furthermore, Dr Rohan Teasdale has made a very useful reference database for this purpose. It is called 'LOCATE' (http://locate.imb.uq.edu.au/), which used the FANTOM3 Isoform Protein Sequence set (Fink *et al.*, 2006; Sprenger *et al.*, 2008). Dr Stefan Wiemann maintains a database called 'LifeDB' (http://www.lifedb.de/lifedb/), which contains expression information about GFP-fused recombinant proteins (Simpson *et al.*, 2000). From such databases and published results, you can confirm the validity of your experiments using the recombinant proteins that you have made. A word of caution is that when you over-express the gene of interest using a very strong promoter [e.g. cytomegalovirus (CMV)], you might need to titer the concentration of plasmid DNA to be transfected. Furthermore, you should establish a 'stable' line [compared with a 'transient' transfection, in which the plasmid (gene) introduced is expressed only in the cell that was originally introduced and can be lost from the cell at any time depending on environmental factors so that the daughter cells will not retain the gene of interest], to obtain consistent results for each assay.

5.3.2.2 Antibody-based validations

One of the most informative methods (but not entirely reliable, unless tested vigorously) is to use an antibody against the protein that your gene of interest encodes. Fortunately, there are international efforts to generate large collections of antibodies, namely the Human Protein Atlas (http://www.proteinatlas.org/) (Uhlén *et al*., 2005). The objective of this database is to provide subcellular localization and expression data of available antibodies for all major human tissues and organs, including 20 different types of cancer. The current version (version 7) of this database is targeting more than 10,100 human genes (~50% of the human protein-coding genes) using 13,150 antibodies (Uhlen *et al*., 2010). As nearly half of the protein-coding genes are covered, you may be surprised to find many antibodies are available for function-unknown genes. In the case of evolutionarily conserved genes, the antibody raised against human genes might be comparable with other species due to sequence similarities. Information relating to the antibodies used for the creation of this database is provided together with a community-based portal called 'Antibodypedia' (http://www.antibodypedia.org/), which lists providers for an antibody of interest. These international consortiums should facilitate efforts of the research community.

To check the specificity of an antibody, you can utilize the recombinant proteins that you made in section 5.3.2.1. You can do this by over-expressing the recombinant proteins in cells (e.g. HEK293 cells) to perform immunohistochemistry (co-localization) and Western blotting experiments. For this, it might be useful to use a cell line in which your gene of interest is not expressed (described in section 5.3.1.1) so that no endogenous protein of interest is detected to minimize the signals. If your antibody passes both tests, you should optimize the condition (e.g. fixation) so that the antibody can be used to detect the endogenous protein of interest in cell lines, tissues and/or embryos. To optimize the conditions under which a new antibody works, the following is the strategy we use for testing it on sections of an adult murine heart:

1. Sacrifice the mouse.
2. Take out the heart and wash it thoroughly in phosphate-buffered saline (PBS) solution.
3. Place the heart on a tissue to wipe off excess solution.
4. Immerse it directly in liquid nitrogen to snap-freeze the tissue. (The tissue can then be stored at –80 °C.)
5. Fix the sample to the metallic cold plate.
   - We prefer using Tissue-Tek O.C.T. Compound (Sakura Finetek, Catalog No. #4583).
   - Instead of using a Cryomold, which is a uniformly shaped flat surface specimen block of plastic, you can simply put a drop of O.C.T. on top of the cold plate and place the frozen heart in the orientation of interest. Then, quickly immerse the whole plate with the sample in liquid nitrogen. Once frozen, you can cover the sample with O.C.T. This saves a lot of O.C.T.
6. Using a cryostat, cryo-section the heart (6-μm thickness).
7. The following four fixations are tested:
   - 4% paraformaldehyde (PFA) at room temperature for 5 min
   - 1% PFA at room temperature for 20 min
   - Aceton at –20 °C for 10 min
   - Aceton/methanol (1:1 mixture) at –20 °C or 10 min
8. Proceed to primary and secondary antibody staining.

Depending on the antibody, antigen retrieval must be performed. More information about antigen retrieval can be found at: http://www.ihcworld.com/_technical_tips/antigen_retrieval_tips.htm. This information is provided by IHC World, LLC. It maintains an excellent collection of protocols for immunohistochemistry (http://www.ihcworld.com/index.htm). Regarding staining of embryos, an excellent protocol for whole-mount embryo staining is provided by Joyner and Wall (2008).

5.4 Step 2: phenotypic analysis

In the field of Developmental Biology, the gold standard is to manipulate (e.g. loss-of-function, gain-of-function) the gene of interest to observe the phenotypes and understand its function in relation to the development of an organism. In the case of Cell Biology, similar methods are used to observe phenotypes by altering the functions of the gene of interest *in vitro* (cell culture). Compared with working with a living organism, experiments performed with cells are faster (not necessarily easier) and cleaner due to fewer cell types in the system. Working with embryonic stem (ES) cells provides knowledge in both fields of studies. This is the reason why many researchers, including us, employ ES cells in their studies. In the following subsections, I describe the techniques currently employed in my group.

5.4.1 Zebrafish as a model organism

The zebrafish (*Danio rerio*; Taxonomy ID: 7955) is a tropical freshwater fish and is regarded as a useful model organism to study vertebrate development and gene function. The zebrafish community is very active in providing information through the Zebrafish Model Organism Database (ZFIN) (http://zfin.org/), including an informative and useful online version of the *Zebrafish Book* (http://zfin.org/zf_info/zfbook/zfbk.html). This book offers experimental protocols and tips for various methods. Straightfoward information about zebrafish is also available from: http://www.neuro.uoregon.edu/k12/zfk12.html. Below is a list of the advantages of using zebrafish as a model organism:

- The genome of zebrafish has been sequenced through the *Danio rerio* Sequencing Project (http://www.sanger.ac.uk/Projects/D_rerio/), which used the strain of zebrafish called 'Tübingen'. Details about this line can be found at: http://zfin.org/action/genotype/genotype-detail?zdbID=ZDB-GENO-010924-10.
- The development of all major organs is extremely rapid (within three days after eggs are laid, which is called the 'hatching

period'). The stages of zebrafish development are recorded in hours, instead of days for mouse. Information about each stage of development can be found at: http://zfin.org/zf_info/zfbook/stages/stages.html.

- The transparent eggs that develop outside the mother's body. Females do not need to be sacrificed to obtain embryos. Furthermore, unlike chicken eggs, live monitoring of embryonic development is possible without further manipulation.
- A sexually mature female zebrafish can lay hundreds of eggs in one breeding attempt, which allows for more specimens to be studied.
- Easy access to *in vivo* embryological manipulations.

One of the key features of using zebrafish as a model organism is great imaging during embryonic development. You can find excellent images of zebrafish embryos in many articles and websites of zebrafish researchers. For example, great images are posted at the websites of Dr Salim Seyfried (http://www.mdc-berlin.de/en/research/research_teams/cell_polarity_and_epithelial_development/Gallery/index.html), the University College London (http://www.ucl.ac.uk/zebrafish-group/research/research.php), Life Technologies (http://tools.invitrogen.com/search/index.cfm?fuseaction=search.simplesearch&navFilter=category%3AGallery&searchterm=zebrafish&selection=-+select+a+gallery+area+-) and Nokon Instruments, Inc. (http://www.nikonsmallworld.com/gallery/search/all/zebrafish) to name just a few.

Disadvantages compared with other model organisms are as follows:

- Zebrafish reach sexual maturity at 3–4 months, longer than for mouse (2 months).
- Although zebrafish are small, they do live in water and require great care to house them. They are sensitive animals, requiring a well-controlled environment, including defined light cycle (14/10-h light–dark cycle).
- The availability of antibodies is rather limited. You can find most of the antibodies used in zebrafish through the ZFIN: http://zfin.org/action/antibody/search.

- When working with the embryos of zebrafish, the materials you obtain (e.g. RNA, proteins) are very limited. For example, when working with 2-day-old embryos, you need more than 20 zebrafish hearts to obtain enough RNA to perform one microarray experiment.
- The zebrafish cannot be inbred; they must be outbred. This results in genetic variations. The *Danio rerio* Sequencing Project used the Tübingen strain for sequencing of the genome, so the sequence information may not apply to the particular strain that you are using. This might pose a problem when designing a primer set for RT-PCR experiments and other types of experiments that require accurate sequence information.

As with any other model organism, there are advantages and disadvantages of using zebafish. In my group, we use zebrafish to look at the question of whether evolutionarily conserved, tissue-enriched, uncharacterized genes are important in the development of our target organ by utilizing a loss-of-function technology. There are different methods available to knockdown the gene of interest. We are using an antisense morpholino-based knockdown method.

Morpholino oligomers are nucleic acid analogs that block access of other molecules (e.g. RNA polymerase, spliceosomes) to targeted nucleic acid sequences (Summerton and Weller, 1991, 1997). Instead of degrading the target RNA molecule, a morpholino binds to the target RNA sequence to interfere with the molecules normally interacting with the RNA (e.g. RNA polymerase, splicing machinery). There are two types of morpholinos: when a morpholino is designed to target the start codon (ATG) for the translation of mRNA to the corresponding protein, the morpholino blocks access of the ribosomal initiation complex so that no protein product of the target gene will be made. This is called a 'translational blocking morpholino (or oligo)'. Using a translational blocking morpholino, no protein that the target gene encodes will be made. Another type of morpholino is called 'splice blocking morpholino (or oligo)'. A splice blocking morpholino is designed at the exon and intron boundary sequence to block access of a small nuclear ribonucleoprotein complex to pre-mRNA to prevent the formation of the splice lariat structure. This causes mis-splicing, which in some cases results in the inclusion of the corresponding intron, which might include a stop codon. As a result, a mis-spliced protein will be made. Unlike a translational blocking morpholino, which can only monitor its effect at the protein level (e.g. Western blotting experiment), the effect of a splice blocking morpholino can be checked by performing RT-PCR to observe a band shift in the products, which are caused by a mis-splicing event. If you would like to know more about morpholinos, there is an excellent review article about gene knockdown by morpholinos, which you can read free at: http://dev.biologists.org/content/135/10/1735.long (Eisen and Smith, 2008). Furthermore, there are informational databases about morpholinos, such as the 'MOrpholino DataBase (MODB)' (http://www.morpholinodatabase.org/). You can also access ZFIN to obtain phenotypic descriptions (including images for some genes) of morpholino-knockdown zebrafish embryos. Because it is difficult to describe the whole procedure of morpholino injection into zebrafish embryos, readers are referred to the excellent movie protocol showing the detailed procedure for performing this type of experiment: http://www.jove.com/details.php?id=1113. If you would like to work with morpholinos, you will need to purchase them from Gene Tools, LLC (http://www.gene-tools.com/). Gene Tools provides a free morpholino design service (http://www.gene-tools.com/Oligo_Design) and excellent support, making the ordering of morpholinos easier.

One of the problems associated with morpholino-based knockdown is its undirected diffusion into the whole egg. This can be solved by using transgenic technology to direct the manipulation of the gene function in a tissue-specific manner, namely via the Cre-loxP system (Sauer, 1987; Sauer and Henderson, 1998). This technology has been used in zebrafish (Speck and Iruela-Arispe, 2009). However, compared with that of mouse, the availability of specific Cre drivers is rather limited. Therefore, we must cope with the problem of undirected diffusion. It should be noted that this type of diffusion is the basis for the injection-based knockdown method. Conceptually, because we are interested in tissue-enriched genes (in other words, the gene of interest is expressed almost exclusively or highly in the organ of interest but not in others), the random diffusion of a morpholino into the whole egg and embryo does not matter as the function of the gene of interest is expressed in a tissue-specific manner. Therefore, in principle, only in the target organ in which our gene is expressed should the phenotypes arise; note that this concept stands only if there is no off-target effect. By accepting this concept, we could solve the problem of the uncontrollable diffusion of a morpholino. Additionally, we have a specific target organ in which to look for the phenotype from the beginning of the experiment. This allows for rapid screening of phenotypes as well as detection of non-specific effects of a morpholino in the organs in which our gene is not expressed highly or is not expressed at all. However, the issue of off-target effects is difficult to deal with, and a number of careful measures must be taken to 'prove' that the morpholino that is employed is acting only on the target gene but not others. To be sure that this concept applies, we check the expression patterns of the gene of interest by inspecting the WISH images provided by the ZFIN (http://zfin.org/cgi-bin/webdriver?MIval=aa-xpatselect.apg), which can also be accessed gene-by-gene, if applicable. Of course, it is better to perform WISH ourselves, but our study focus is *in silico* screening, which points to the use of such information over the internet.

One of the gold standards in experiments using morpholinos to rule out potential off-target effects (reviewed extensively by Eisen and Smith, 2008) is to rescue the phenotype by over-expression of the target gene (Cibois *et al.*, 2010). This can be achieved by an injection of the mRNA of the target gene. Rescue by such means poses a problem in our approach as our target gene is expressed in a tissue-specific manner. As in the case of morpholinos, an injection of RNA results in random diffusion throughout the egg and embryo. Because our target gene is expressed in a certain set of tissues but not others (or is expressed only to low level in them), this results in a gain-of-function in our case; in other words, phenotypes may arise due to the introduction of our target gene and its protein product in organs and tissues that do not require them for normal development. Thus, in our case, this gold standard may not necessarily apply.

Given the above facts and concepts, we employ morphlino-based screening of gene functions using zebrafish embryos as an initial functional characterization of evolutionarily conserved, tissue-enriched, uncharacterized genes. In studying the development of heart, zebrafish is a superb model organism because its heart develops within 48 h after the egg is laid, and gas exchange of an embryo does not rely on oxygen transport by blood (cutaneous gas exchange) (Barrionuevo and Burggren, 1999). The latter is useful because although an embryo might have a heart function defect due to injection of a morpholino targeting the gene of interest, it can still develop up to a week after birth without dying. Needless to say, this concept stands if and only if there are no major developmental defects in any other organs, which is typical if a high concentration of morpholino is used and if off-target effects are present. Because the embryo develops externally, the phenotypes can be monitored in real time. This is a distinct advantage of using zebrafish because we can perform live imaging of the phenotypes of an embryo to record how the loss-of-function assay for the target gene affects the development of its heart in a time-dependent manner. Because the embryo is small [1.9 mm at 24 h post-fertilization (hpf), 3.1 mm at 48 hpf and 3.5 mm at 3 days post-fertilization (data taken from the Zebrafish Atlas, http://zfatlas.psu.edu/, NIH grant 5R24 RR01744, Jake Gittlen Cancer Research Foundation, and PA Tobacco Settlement Fund) (http://zfatlas.psu.edu/progress.php)], this allows for four-dimensional imaging when a GFP line of the zebrafish is used without sacrificing and sectioning the embryo (Yu *et al.*, 2006). However, to obtain a good image, the embryo needs to be sectioned. There is an excellent protocol available for this (Sullivan-Brown *et al.*, 2011).

As mentioned above, zebrafish offers a great opportunity to screen for genes whose functions might be necessary for the development of a target organ. Given that it is a live organism, zebrafish offers a fast system to obtain phenotypes *in vivo* in less than 1 week. Compared with the mouse system, with which you need first to make the founder line and cross them further to obtain homozygous mice to check for phenotypes (which can take over a year), you can perform a loss-of-function experiment by ordering a morpholino and injecting it into eggs in less than 2 weeks. This significantly reduces the time required to determine whether the gene of interest is important for the development of the target organ. Although zebrafish offers great potential in the field of Developmental Biology, conducting biochemical assays using morpholino-based loss-of-function experiments can be problematic as the results may vary due to the degrees of knockdown, which differ among the injected embryos. Furthermore, the amount of tissue that can be obtained from one embryo is very limited, which makes it difficult to perform biochemical assays. We therefore use *in vitro* culture to study the functions of our target genes, as described in the next section.

5.4.2 *In vitro* assays

Conceptually, compared with the *in vivo* system described above, *in vitro* experiments are more controllable (though not necessarily easier). Most cells can proliferate and divide under controlled conditions. Unlike primary cells, cells of an established or immortalized cell line can proliferate indefinitely, allowing experiments to be performed continuously. As the 'art of cell culturing' is a field in itself, I will not cover it in this chapter; rather, I will explain the methods that we are currently employing to manipulate the functions (most genes have more than one function, depending on the biological situation) of the gene of interest.

ES cells offer great opportunities and advantages over other cell lines in studying the development of tissues. ES cells are pluripotent [able to form all the body's cell lineages (Smith, 2006)] stem cells that are derived from the inner cell mass of the blastocyst (Evans and Kaufman, 1981). We work with murine ES cells for the following reasons:

- They expand indefinitely and are easy to culture.
- They possess an ability to differentiate into different cell types, including cardiomyocytes, which is our primary interest.
- They are easy to manipulate by gain- and loss-of-functional methods.

The second point, which is called 'pluripotency', is a double-edged sword because it is hard to obtain a homogeneous cell type during cell culturing. To circumvent this problem, we have introduced a selection construct with a neomycin-resistance gene cassette that is placed under the promoter of *Myh6* to select for only those cells differentiated into cardiomyocytes (Klug *et al.*, 1996; Gellert *et al.*, 2012). As originally established by Dr Robert Zweigerdt and others (Zandstra *et al.*, 2003; Zweigerdt *et al.*, 2003), the rotation culturing method is used to induce the differentiation of ES cells into beating cardiomyocyotes, instead of the hanging drop technique or induction by drugs and molecules (e.g. retinoic acids) (Yamashita, 2010). To select for cardiomyocytes, G418 (also known as Geneticin®), which is an analog of neomycin sulfate (http://www.invitrogen.com/site/us/en/home/Products-and-Services/Applications/Cell-Culture/Transfection/Selection/g418.html), is administered to the culture medium on day 9 of differentiation, which coincides with the expression of *Myh6* at a high level. In principle, those cells that do not express *Myh6* (e.g. endothelial cells, fibroblasts) will be eliminated from the culture. One cautionary note is that these differentiated cardiomyocytes differ significantly from mature adult cardiomyocytes physiologically as well as at the levels of gene and protein expression (Fijnvandraat *et al.*, 2003; Gellert *et al.*, 2012). A similar method to purify the population of differentiated cells has been used by others for cardiomyocytes using the early cardiac marker *Nkx-2.5* (Hsiao *et al.*, 2008), for endothelial cells through the use of *tie-1* (Marchetti *et al.*, 2002) and for neural cells by *Sox2* (Li *et al.*, 1998) to name but a few.

Having such a convenient culture system, you can manipulate the function of the gene of interest by over-expressing the fusion protein (see section 5.3.2.1) in a gain-of-function experiment or by down-regulating the expression of the target gene by a loss-of-function assay. There are different methods to achieve the knockdown of the gene of interest, including RNA interference (RNAi) (Fire *et al.*, 1998). RNAi technology is a natural process in which the activity of a target gene is controlled through the formation of an RNA-interfering silencing complex (RISC) together with one of two types of small RNA molecules, namely microRNA (miRNA) and small interfering RNA (siRNA). miRNA suppresses the translation of the target gene, whereas siRNA interferes with post-transcriptional processes by binding to the specific mRNA sequence of the target gene (Rao *et al.*, 2009).

As this is a powerful tool, laboratories around the world are using this technology to perform high-throughput screening using genome-wide RNAi libraries (Falschlehner *et al.*, 2010). Such an approach meets nicely with our screening demands; thus, we prefer to use lentivirus-based short hairpin RNA (shRNA) knockdown, which works efficiently at the levels of infection, as well as knockdown of a target gene (Manjunath *et al.*, 2009), to perform functional studies of our target genes. Like morpholinos described previously, off-target effects of shRNA are a major problem. To check for the validity of shRNAs, it is typically suggested to use several siRNAs targeting the gene of interest as well as the use of a proper control, such as an empty viral vector or irrelevant shRNA cassette (e.g. scrambled, no-target or mutated control) (Singer and Verma, 2008). As such a system is commercially available [e.g. MISSION® (Sigma-Aldrich Co. LLC.), shRNA and siRNA Lentiviral Cloning and Expression System (System Biosciences, Inc.)], it will not be discussed here. Please note that working with a virus means that all materials must be treated with appropriate biosafety precautions.

With the methodology explained above, you have a reasonably pure culture system for a defined cell lineage and cell type to study the functions of a gene of interest to perform further functional studies by manipulating the target gene. The readouts could be by RT-PCR experiments for tissue-specific marker genes (e.g. *Myh6* for cardiomyocytes). In addition, they could be provided by immunocytochemistry and flow cytometry analysis using antibodies against tissue-specific markers. Needless to say, the cells obtained could be used for transcriptomics (e.g. microarrays) as well as proteomics (e.g. mass spectrometry) to screen for affected signaling pathways.

5.4.3 Mouse as a model organism

Do you remember who was awarded the Nobel Prize in Physiology or Medicine for 2007? You might if you are a developmental biologist. The Nobel Laureates were Drs Mario R. Capecchi, Martin J. Evans and Oliver Smithies for their discoveries of the ‘principles for introducing specific gene modifications in mice by the use of embryonic stem cells' (The 2007 Nobel Prize in Physiology or Medicine – Press Release. Nobelprize.org. 11 September 2011, http://www.nobelprize.org/nobel_prizes/medicine/laureates/2007/press.html). Their discovery led to the creation of a technology called 'gene targeting' in mice. We can benefit greatly from this technology in the field of Developmental Biology, by inactivating the functions of one gene to create a knockout (KO) mouse line. A detailed description of how to make a KO mouse line goes well beyond the scope of this book. I will describe the strategy we are currently employing to further extend our 1,000 Genes Project through conditional KO mice.

If a gene of interest is a critical one that is required for development of a mouse, you might see severe phenotypes in which an embryo might die during development. It is estimated that 30% of KO homozygous mouse lines result in embryonic or early postnatal lethality (Kühn and Schwenk, 2003). Unlike zebrafish embryos, murine embryos develop in the body of their mother. Therefore, you cannot analyze the phenotypes unless the mother is sacrificed. Furthermore, a problem associated with the embryonic lethality of KO mice is that it is hard to determine the primary cause of death as you will see effects secondary to the primary cause of the phenotypes. To overcome this problem, researchers use conditional KO mice.

The conditional knockout strategy allows for a condition-specific inactivation of the gene of interest (e.g. tissue-specific). This is achieved by the expression of a site-specific DNA recombinase [Cre (Sauer and Henderson, 1998) or Flippase (FLP) (Zhu and Sadowski, 1995) recombination enzyme] in mice, which are crossed with mice bearing the recombinase-specific sites in the non-coding regions of the target gene: loxP for Cre-recombinase or Flippase Recognition Target (FRT) for FLP-recombinase (Kühn and Schwenk, 2003). Of the above two systems, we use the Cre-loxP system. A Cre-recombinase-bearing mouse line is called 'cre driver', for which there is a comprehensive informational database from the Mouse Genome Informatics (MGI): http://www.informatics.jax.org/recombinase.shtml. As for the mouse line with loxP sites, which are collectively called 'conditional alleles', there are international efforts towards

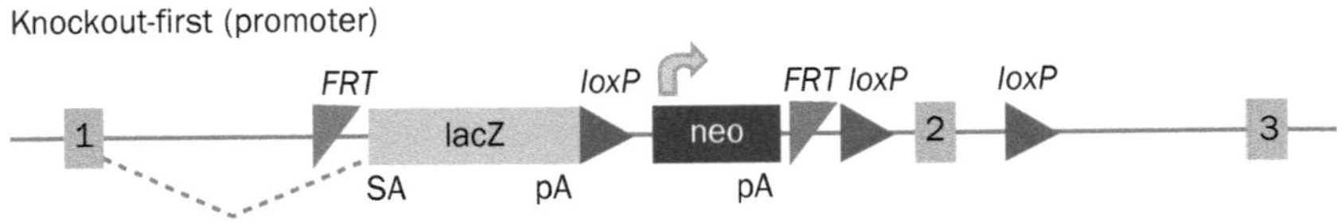

Figure 5.2 Knockout-first (promoter) allele. The targeted exon is 'Exon 2', which is called 'critical exon' by the International Knockout Mouse Consortium. Reproduced from the International Knockout Mouse Consortium Web Portal (www.knockoutmouse.org) with permission of the copyright holders, The Jackson Laboratory and the Wellcome Trust Sanger Institute

large-scale mouse knockout (Guan *et al.*, 2010) organized by the International Knockout Mouse Consortium (IKMC), which aims to mutate all protein-coding genes in mouse using gene trapping and gene targeting in C57BL/6 ES cells (http://www.knockoutmouse.org/). By accessing this website, you can search for the presence of the conditional allele for a gene of interest. For details, please refer to the above website. If the gene of interest is not listed, you can request the gene be knocked via: http://www.knockoutmouse.org/startnominategene. We are using the following strategy to study the functions of our genes:

1. Screen for a founder line. For this, we utilize the WISH images that we obtained previously using E9.5 and 10.5 murine embryos (Uchida *et al.*, 2009) (see section 5.3.1.2 for details). In the case of the 'Knockout-first (promoter)' allele (Testa *et al.*, 2004), the LacZ gene is inserted as shown Figure 5.2.

By performing LacZ staining with the following protocol, we can compare the staining pattern of WISH with that of LacZ staining of embryos obtained by crossing a chimera with a wild-type mouse (e.g. C57BL/6). This allows us to screen for a founder line with the expected expressions.

Protocol for LacZ staining of embryos

(1) Fix an embryo with the following fixation solution at room temperature for 5 min:

To prepare for 10 mL of fixation solution	
Glutaraldehyde (25% stock solution)	160 μL
$MgCl_2$ (1 M)	20 μL
EGTA (0.5 M, pH 8.0)	100 μL
Combine up to 10 mL with 1 × PBS	

Note: Any form of formaldehyde [e.g. paraformaldehyde (PFA)] will kill the signal of LacZ.

(2) Wash three times with 1 × PBS at room temperature for 5 min each wash.

(3) Add the following LacZ staining solution to the embryo.

To prepare for 10 mL of LacZ staining solution	
$MgCl_2$ (1 M)	20 μL
NP40 (10%)	200 μL
Na-desoxycholate (1%)	1 mL
K_3FeCN_6 (500 mM)	100 μL
K_4FeCN_6 (500 mM)	100 μL
X-Gal (20 mg/mL stock)	500 μL
Combine up to 10 ml with 1 × PBS	

(4) Stain the embryo overnight at 37 °C.

2. After a founder line is screened, the following is the strategy recommended to produce reporter knockouts, conditional knockouts and null alleles using Cre and FLP-recombinases (Figure 5.3).

For this purpose, we use the following breeding scheme with F1 generation mice. (a) To make the tm1b allele, breed with a general Cre-deleter mouse line [e.g. Tg(CMV-cre)1Cgn], which carries Cre-recombinase under a human cytomegalovirus (CMV) promoter to express Cre-recombinase ubiquitously in most cells and tissues (Schwenk *et al.*, 1995) (http://www.informatics.jax.org/javawi2/servlet/WIFetch?page=alleleDetail&key=6210#imsr). Conceptually, because the target genes that we selected are expressed in a tissue-enriched or tissue-specific manner, using the ubiquitous Cre line above

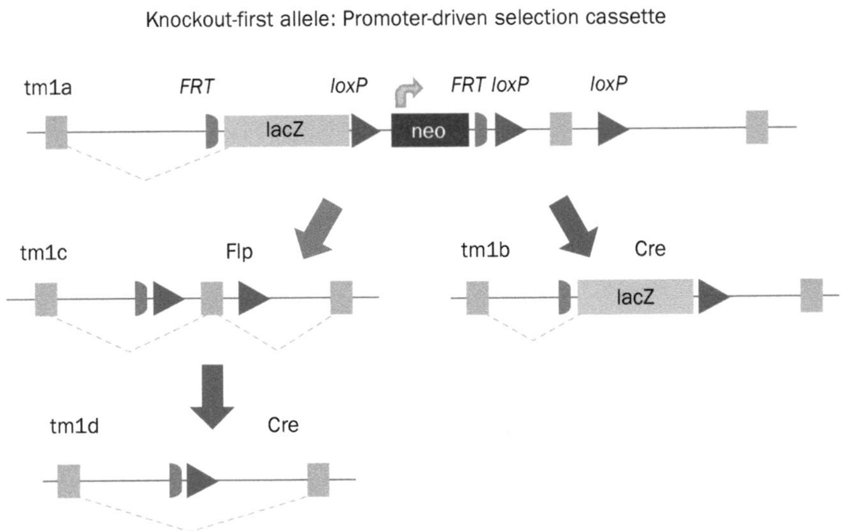

Figure 5.3 The strategy proposed by the International Knockout Mouse Consortium. The terminologies used are: 'tm1a', KO first allele (reporter-tagged insertion allele); 'tm1b', post-Cre (reporter-tagged deletion alllele); 'tm1c', post-Flp (conditional allele); and 'tm1d', post-Flp and Cre (deletion allele with no reporter). Reproduced from the International Knockout Mouse Consortium Web Portal (www.knockoutmouse.org) with permission of the copyright holders, The Jackson Laboratory and the Wellcome Trust Sanger Institute

should not interfere with the tissues with no or much less expressed tissues and organs. (b) To produce the tm1c allele, cross with a general FLP deletion mouse line [e.g. Gt(ROSA)26Sor$^{tm1(FLP1)Dym}$], which carries FLP-recombinase under the control of the constitutive ROSA26 locus to achieve expression of most tissue types, including cells of the developing germ lines (Farley *et al.*, 2000) (http://www.informatics.jax.org/javawi2/servlet/WIFetch?page=alleleDetail&key=9406). It is recommended to cross at least one more generation to completely restore the wild-type expression of the target gene by removing the trapping cassette. For details, please refer to: http://www.knockout mouse.org/kb/entry/104/ and http://www.knockoutmouse.org/kb/entry/103/

As the breeding of mice can take considerable time (from putting male and female mice together to the sexually matured next-generation mice for breeding will take at least 3 months), we are currently crossing an F1 male mouse of the target gene to the following three female mice at one breeding: C57BL/6 (to maintain the mouse line); Tg(CMV-cre)1Cgn (to quickly look at the phenotypes of ablating the gene of interest); and Gt(ROSA)26Sor$^{tm1(FLP1)Dym}$ (to restore the wild-type expression of the target gene for further studies).

3. If phenotypes arise, proceed with analysis and consider employing more tissue-specific Cre drivers, which can be identified from the MGI website (http://www.informatics.jax.org/recombinase.shtml).

Recently, the Sanger Institute introduced the miRNA knockout (mirKO) resource, which can be searched through the IKMC website (http://www.knockoutmouse.org/) (Prosser *et al.*, 2011). Of note, all the KO mice are linked by the IKMC MartSearch project (http://www.knockoutmouse.org/martsearch/), which allows *in silico* screening as described in Chapter 4.

As stated above, the advantage of using knockout mice is complete ablation of the target gene by genetic means, which allows us to study its functions under the 'null' condition. Of course, working with knockout mice takes a long time. From the creation of a targeting vector to having a colony of homozygous mice to be analyzed, it will take well over a year, depending on your efforts, and simply luck; after all, you need to screen for a founder line that you can work with. Therefore, before we commit ourselves, we employ antisense morphlino-based knockdown in zebrafish embryos as described in section 5.4.1 to have an additional layer of screening.

5.5 Step 3: molecular interactions

Having obtained the phenotypes in which the gene of interest is expressed, the next obvious question is what is the mechanism by which such a phenotype arises due to the impairment of the function of your gene of interest? In fact, this is one of the most time-consuming and challenging questions in Biology. However, *in silico* screening can be utilized to make an educated guess about the functions of the gene of interest. Dr Ben Lehner published an interesting study in 2009 (Bossi, 2009), in which the authors concluded that most tissue-specific proteins bind to universally expressed proteins, whereas most 'housekeeping' proteins (a group of universally expressed proteins) interact with highly tissue-specific proteins. Accepting this concept, for tissue-enriched genes, protein–protein interactions (PPIs) might occur with universally expressed (ubiquitous) proteins. Conceptually, this is convenient because most of the large-scale screening for PPI often uses ubiquitous genes/proteins as baits to screen for binding partners. The datasets and databases built from such large-scale screenings are listed in sections 2.3.2.4.4 and 2.4.3 as well as in Dr Lehner's article (Bossi, 2009). Among them, the Genome Network Platform (http://genomenetwork.nig.ac.jp/index_e.html) is particularly useful.

The development of high-end mass spectrometry has truly opened up a new horizon in the field of Biology. Although the price of such machines is still beyond the budget of a single lab, more and more research institutes and universities are installing mass spectrometry machines to cope with the increasing need to profile biological phenomena at the protein level. If you have already made recombinant proteins as described in section 5.3.2.1, you could utilize them to perform a pull-down (immunoprecipitation) assay to screen for binding partners. Of course, if an antibody against the protein that your gene of interest encodes works for immunoprecipitation, then you could also pull down the protein complex made by the endogenous protein of interest. As there are many excellent protocols and products available and the technology itself is evolving quite rapidly, I merely refer readers to the following useful sites for detailed information:

- Abcam: http://www.abcam.com/index.html?pageconfig=resource&rid=11385
- Cell Signaling Technology: http://www.cellsignal.com/support/protocols/IP_Native_Protein.html
- Millipore: http://www.millipore.com/immunodetection/id3/immunoprecipitationprotocols
- Promega: http://www.promega.com/resources/product-guides-and-selectors/protocols-and-applications-guide/protein-purification-and-analysis/

- Protocol Online: http://www.protocol-online.org/prot/Molecular_Biology/Protein/Immunoprecipitation__IP_/index.html
- Sigma-Aldrich: http://www.sigmaaldrich.com/life-science/cell-biology/antibodies/antibodies-application/protocols/immunoprecipitation.html
- Thermo Scientific: http://www.piercenet.com/browse.cfm?fldID=F3FD3612-415F-42A5-8922-736F9FDD36FB

As one could easily imagine, current proteomics technology lags behind that of transcriptomics. There are two main reasons for this: (1) a large amount of materials are needed to profile as no method exists to amplify the proteins contained in the starting materials, whereas PCR amplifications are available for transcriptomics; and (2) there is limited resolution to the identified peptides due to the use of a restricted amount of peptidases. Unlike transcriptomics, which can be designed to separate similar mRNA sequences by targeting the unmatched sequences, the differences between similar amino acid sequences cannot be differentiated if there are no peptidases that cut the similar sequences and leave the difference to be read. This is the reason why many mass spectrometry peptide datasets are mapped to more than one protein ('protein groups') instead of showing a one-to-one relationship between a peptide and protein (Uchida *et al.*, 2011). Although the SILAC (stable isotope labeling using amino acids in cell culture) method is currently favored for quantitative proteomics to detect differences in protein abundance among samples using non-radioactive isotopic labeling (Ong *et al.*, 2003; Ong and Mann, 2006), the problem of limited resolution persists. However, with SILAC, the number of peptides can be reduced significantly by performing a comparative method to reduce false-positive bindings. Although there are still many technical challenges for mass-spectrometry-based proteomics, the waves of PPI datasets generated by this approach will greatly help us to further screen for binding partners *in silico*. Therefore, technical advances in the field proteomics will be of great interest.

Whether based on *in silico* screening or actual proteomics experiments using mass spectrometry, PPI results must be validated by further biological experiments to provide ultimate confirmation of the pull down of endogenous binding partners using the antibody against the target protein. However, this is very difficult to do. More common practice is to over-express both the protein that your gene of interest encodes and its potential binding partner using different tags (e.g. GFP vs. His). For this, we recommend using μMACS™ Epitope Tag Isolation Kits (http://www.miltenyibiotec.com/en/NN_162_Isolation_and_analysis_of_epitope_tagged_proteins.aspx), which utilize magnetic microbeads. This kit is easy to use and the detailed protocol is provided by Miltenyi Biotec. To detect the interactions, the eluted immunoprecipitated protein complexes can be analyzed by Western blotting using an appropriate antibody.

Jain *et al.* (2011) have described a novel method, the 'single-molecule pull-down (SiMPull)' assay, that allows PPIs to be studied at the single-molecule level using fluorescence microscopy. As commented by Tinnefeld (2011), SiMPull can be considered as another version of a Western blotting assay. It will still be some time before this method and similar single-molecule assays become routine, but it is something that you should keep in mind for further validation experiments.

5.6 Other experiments

In the above sections, some of the typical experiments that you need to perform to validate the results of your *in silico* screening and predictions are provided. As mentioned at the beginning of this chapter, the types of experiments to be performed vary case-by-case (gene-by-gene). For example, if you are interested in working with a gene with DNA-binding property (e.g. transcription factors), you will need to find the binding sequence and sites of the gene of interest using methods such as the electrophoresis mobility shift assay (EMSA) (Hellman and Fried, 2007), ChIP–chip or ChIP–seq (Farnham, 2009). As the technology has advanced in recent years, attention has shifted more towards 'single-cell analysis'. More and more studies indicate that each cell behaves differently at the levels of gene and protein expression even when derived from a common tissue source or progenitor. A recent study by Dr Pamela Silver indicates that even cells of the most commonly used line, HeLa, have alternative isoform differences among them (Waks *et al.*, 2011). A typical mammalian cell contains ~10 pg total RNA and ~0.1 pg mRNA (Tang, 2011). Although the technology has advanced, analyzing such a limited amount of RNA to obtain a global expression profile of a single cell remains a major challenge (Gonzalez-Roca *et al.*, 2010; Reiter *et al.*, 2011), but not impossible, as demonstrated by Drs Kazuki Kurimoto and Mitinori Saitou (Kurimoto *et al.*, 2007; Kawaguchi *et al.*, 2008; Kurimoto and Saito, 2010, 2011; Hayashi *et al.*, 2011). We can expect more datasets of single-cell analysis in the near future.

5.7 Conclusion

In this chapter I have described typical experiments that should be performed to validate one's *in silico* screening and predition. Unfortunately, I cannot give general guidelines for the types of experiments to be performed further, simply because this will vary case-by-case. When planning for biological experiments, you should first check whether genes similar to your gene of interest have been studied. You can simply follow what others have done to show the functionalities of such genes. Of course, each gene is different, and as such you will need to modify the experiments and further develop methods to provide sound results. To put it simply: your imagination is the key to success.

References

Barrionuevo WR, Burggren WW. (1999) O_2 consumption and heart rate in developing zebrafish (*Danio rerio*): influence of temperature and ambient O_2. *Am J Physiol* 276(2): R505–13.

Bossi A, Lehner B. (2009) Tissue specificity and the human protein interaction network. *Mol Syst Biol* 5: 260.

Bustin SA. (2000) Absolute quantification of mRNA using real-time reverse transcription polymerase chain reaction assays. *J Mol Endocrinol* 25(2): 169–93.

Cibois M, Gautier-Courteille C, Vallée A, Paillard L. (2010) A strategy to analyze the phenotypic consequences of inhibiting the association of an RNA-binding protein with a specific RNA. *RNA* 16(1): 10–15.

Eisen JS, Smith JC. (2008) Controlling morpholino experiments: don't stop making antisense. *Development* 135(10): 1735–43.

Evans MJ, Kaufman MH. (1981) Establishment in culture of pluripotential cells from mouse embryos. *Nature* 292(5819): 154–6.

Falschlehner C, Steinbrink S, Erdmann G, Boutros M. (2010) High-throughput RNAi screening to dissect cellular pathways: a how-to guide. *Biotechnol J* 5(4): 368–76.

Farley FW, Soriano P, Steffen LS, Dymecki SM. (2000) Widespread recombinase expression using FLPeR (flipper) mice. *Genesis* 28(3-4): 106–10.

Farnham PJ. (2009) Insights from genomic profiling of transcription factors. *Nat Rev Genet* 10(9): 605–16.

Fijnvandraat AC, van Ginneken AC, de Boer PA, *et al.* (2003) Cardiomyocytes derived from embryonic stem cells resemble cardiomyocytes of the embryonic heart tube. *Cardiovasc Res* 58(2): 399–409.

Fink JL, Aturaliya RN, Davis MJ, *et al.* (2006) LOCATE: a mouse protein subcellular localization database. *Nucleic Acids Res* 34(Database issue): D213–17.

Fire A, Xu S, Montgomery MK, Kostas SA, Driver SE, Mello CC. (1998) Potent and specific genetic interference by double-stranded RNA in *Caenorhabditis elegans*. *Nature* 391(6669): 806–11.

Gellert P, Jenniches K, Braun T, Uchida S (2010) C-It: a knowledge database for tissue-enriched genes. *Bioinformatics* 26(18): 2328–33.

Gellert P, Teranishi M, Jenniches K, *et al.* (2012) Gene Array Analyzer: alternative usage of gene arrays to study alternative splicing events. *Nucleic Acids Res* 40(6): 2414–25. http://www.ncbi.nlm.nih.gov/pubmed/22123740

Gonzalez-Roca E, Garcia-Albéniz X, Rodriguez-Mulero S, Gomis RR, Kornacker K, Auer H. (2010) Accurate expression profiling of very small cell populations. *PLoS One* 5(12): e14418.

Guan C, Ye C, Yang X, Gao J. (2010) A review of current large-scale mouse knockout efforts. *Genesis* 48(2): 73–85.

Hayashi K, Ohta H, Kurimoto K, Aramaki S, Saitou M. (2011) Reconstitution of the mouse germ cell specification pathway in culture by pluripotent stem cells. *Cell* 146(4): 519–32.

Hellman LM, Fried MG. (2007) Electrophoretic mobility shift assay (EMSA) for detecting protein–nucleic acid interactions. *Nat Protoc* 2(8): 1849–61.

Hsiao EC, Yoshinaga Y, Nguyen TD, *et al.* (2008) Marking embryonic stem cells with a 2A self-cleaving peptide: a NKX2-5 emerald GFP BAC reporter. *PLoS One* 3(7): e2532.

Jain A, Liu R, Ramani B, *et al.* (2011) Probing cellular protein complexes using single-molecule pull-down. *Nature* 473(7348): 484–8.

Joyner A, Wall N. (2008) Immunohistochemistry of whole-mount mouse embryos. CSH Protoc pdb.prot4820. http://www.ncbi.nlm.nih.gov/pubmed/21356665

Kawaguchi A, Ikawa T, Kasukawa T, *et al.* (2008) Single-cell gene profiling defines differential progenitor subclasses in mammalian neurogenesis. *Development* 135(18): 3113–24.

Klug MG, Soonpaa MH, Koh GY, Field LJ. (1996) Genetically selected cardiomyocytes from differentiating embronic stem cells form stable intracardiac grafts. *J Clin Invest* 8(1): 216–24.

Kühn R, Schwenk F. (2003) Conditional knockout mice. *Methods Mol Biol* 209: 159–85.

Kurimoto K, Yabuta Y, Ohinata Y, Saitou M. (2007) Global single-cell cDNA amplification to provide a template for representative high-density oligonucleotide microarray analysis. *Nat Protoc* 2(3): 739–52.

Kurimoto K, Saitou M. (2010) Single-cell cDNA microarray profiling of complex biological processes of differentiation. *Curr Opin Genet Dev* 20(5): 470–7.

Kurimoto K, Saitou M. (2011) A global single-cell cDNA amplification method for quantitative microarray analysis. *Methods Mol Biol* 687: 91–111.

Li M, Pevny L, Lovell-Badge R, Smith A. (1998) Generation of purified neural precursors from embryonic stem cells by lineage selection. *Curr Biol* 8(17): 971–4.

Manjunath N, Wu H, Subramanya S, Shankar P. (2009) Lentiviral delivery of short hairpin RNAs. *Adv Drug Deliv Rev* 61(9): 732–45.

Marchetti S, Gimond C, Iljin K, *et al.* (2002) Endothelial cells genetically selected from differentiating mouse embryonic stem cells incorporate at sites of neovascularization in vivo. *J Cell Sci* 115(10): 2075–85.

Mercer TR, Dinger ME, Sunkin SM, Mehler MF, Mattick JS. (2008) Specific expression of long noncoding RNAs in the mouse brain. *Proc Natl Acad Sci USA* 105(2): 716–21.

Morrison TB, Weis JJ, Wittwer CT. (1998) Quantification of low-copy transcripts by continuous SYBR Green I monitoring during amplification. *Biotechniques* 24: 954–62.

Neilson EG. (2010) The Jeremiah Metzger lecture. The origin of fibroblasts and the terminality of epithelial differentiation. *Trans Am Clin Climatol Assoc* 121: 240–50.

Ong SE, Kratchmarova I, Mann M. (2003) Properties of 13C-substituted arginine in stable isotope labeling by amino acids in cell culture (SILAC). *J Proteome Res* 2(2): 173–81.

Ong SE, Mann M. (2006) A practical recipe for stable isotope labeling by amino acids in cell culture (SILAC). *Nat Protoc* 1(6): 2650–60.

Palmer E, Freeman T. (2004) Investigation into the use of C- and N-terminal GFP fusion proteins for subcellular localization studies using reverse transfection microarrays. *Comp Funct Genomics* 5(4): 342–53.

Prosser HM, Koike-Yusa H, Cooper JD, Law FC, Bradley A. (2011) A resource of vectors and ES cells for targeted deletion of microRNAs in mice. *Nat Biotechnol* 7 August. doi: 10.1038/nbt.1929. [Epub ahead of print]

Rao DD, Vorhies JS, Senzer N, Nemunaitis J. (2009) siRNA vs. shRNA: similarities and differences. *Adv Drug Deliv Rev* 61(9): 746–59.

Reiter M, Kirchner B, Müller H, Holzhauer C, Mann W, Pfaffl MW. (2011) Quantification noise in single cell experiments. *Nucleic Acids Res* 39(18): e124.

Rinn JL, Bondre C, Gladstone HB, Brown PO, Chang HY. (2006) Anatomic demarcation by positional variation in fibroblast gene expression programs. *PLoS Genet* 2(7): e119.

Sanger JW, Kang S, Siebrands CC, *et al.* (2005) How to build a myofibril. *J Muscle Res Cell Motil* 26(6-8): 343–54.

Sauer B. (1987) Functional expression of the cre-lox site-specific recombination system in the yeast *Saccharomyces cerevisiae. Mol Cell Biol* 7(6): 2087–96.

Sauer B, Henderson N. (1998) Site-specific DNA recombination in mammalian cells by the Cre recombinase of bacteriophage P1. *Proc Natl Acad Sci USA* 85(14): 5166–70.

Schwenk F, Baron U, Rajewsky K. (1995) A cre-transgenic mouse strain for the ubiquitous deletion of loxP-flanked gene segments including deletion in germ cells. *Nucleic Acids Res* 23(24): 5080–1.

Simpson JC, Wellenreuther R, Poustka A, Pepperkok R, Wiemann S. (2000) Systematic subcellular localization of novel proteins identified by large-scale cDNA sequencing. *EMBO Rep* 1(3): 287–92.

Singer O, Verma IM. (2008) Applications of lentiviral vectors for shRNA delivery and transgenesis. *Curr Gene Ther* 8(6): 483–8.

Smith, A. (2006) Glossary A glossary for stem-cell biology. *Nature* 441: 1060.

Speck NA, Iruela-Arispe ML. (2009) Conditional Cre/LoxP strategies for the study of hematopoietic stem cell formation. *Blood Cells Mol Dis* 43(1): 6–11.

Sprenger J, Lynn Fink J, Karunaratne S, Hanson K, Hamilton NA, Teasdale RD. (2008) LOCATE: a mammalian protein subcellular localization database. *Nucleic Acids Res* 36(Database issue): D230–3.

Sullivan-Brown J, Bisher ME, Burdine RD. (2011) Embedding, serial sectioning and staining of zebrafish embryos using JB-4 resin. *Nat Protoc* 6(1): 46–55.

Sumanas S, Larson JD. (2002) Morpholino phosphorodiamidate oligonucleotides in zebrafish: a recipe for functional genomics? *Brief Funct Genomic Proteomic* 1(3): 239–56.

Summerton J, Weller D. (1991) *Uncharged morpholinobased polymers having achiral intersubunit linkages.* US Patent 5034506.

Summerton J, Weller D. (1997) Morpholino antisense oligomers: design, preparation, and properties. *Antisense Nucleic Acid Drug Dev* 7(3): 187–95.

Tang F, Lao K, Surani MA. (2011) Development and applications of single-cell transcriptome analysis. *Nat Methods* 8(Suppl. 4): S6–11.

Testa G, Schaft J, van der Hoeven F, *et al.* (2004) A reliable lacZ expression reporter cassette for multipurpose, knockout-first alleles. *Genesis* 38(3): 151–8.

Tinnefeld P. (2011) Protein–protein interactions: pull-down for single molecules. *Nature* 473(7348): 461–2.

Uchida S, Fuke S, Tsukahara T. (2007) Upregulations of Gata4 and oxytocin receptor are important in cardiomyocyte differentiation processes of P19CL6 cells. *J Cell Biochem* 100(3): 629–41.

Uchida S, Schneider A, Wiesnet M, *et al.* (2009) An integrated approach for the systematic identification and characterization of heart-enriched genes with unknown functions. *BMC Genomics* 10: 100.

Uchida S, Gellert P, Braun T. (2011) Deeply dissecting stemness: making sense to non-coding RNAs in stem cells. *Stem Cell Rev* 25 June. [Epub ahead of print] DOI: 10.1007/s12015-011-9294-y

Uhlén M, Björling E, Agaton C, *et al.* (2005) A human protein atlas for normal and cancer tissues based on antibody proteomics. *Mol Cell Proteomics* 4(12): 1920–32.

Uhlen M, Oksvold P, Fagerberg L, *et al.* (2010) Towards a knowledge-based Human Protein Atlas. *Nat Biotechnol* 28(12): 1248–50.

Vandesompele J, De Preter K, Pattyn F, *et al.* (2002) Accurate normalization of real-time quantitative RT-PCR data by geometric averaging of multiple internal control genes. *Genome Biol* 3(7): RESEARCH0034.

VanGuilder HD, Vrana KE, Freeman WM. (2008) Twenty-five years of quantitative PCR for gene expression analysis. *Biotechniques* 44(5): 619–26.

Waks Z, Klein AM, Silver PA. (2011) Cell-to-cell variability of alternative RNA splicing. *Mol Syst Biol* 7: 506.

Yamashita JK. (2010) ES and iPS cell research for cardiovascular regeneration. *Exp Cell Res* 316(16): 2555–9.

Yu RM, Lin CC, Chan PK, *et al.* (2006) Four-dimensional imaging and quantification of gene expression in early developing zebrafish (*Danio rerio*) embryos. *Toxicol Sci* 90(2): 529–38.

Zandstra PW, Bauwens C, Yin T, *et al.* (2003) Scalable production of embryonic stem cell-derived cardiomyocytes. *Tissue Eng* 9(4): 767–78.

Zhu XD, Sadowski PD. (1995) Cleavage-dependent ligation by the FLP recombinase. Characterization of a mutant FLP protein with an alteration in a catalytic amino acid. *J Biol Chem* 270(39): 23044–54.

Zweigerdt R, Burg M, Willbold E, Abts H, Ruediger M. (2003) Generation of confluent cardiomyocyte monolayers derived from embryonic stem cells in suspension: a cell source for new therapies and screening strategies. *Cytotherapy* 5(5): 399–413.

6

Further application: adult stem cells and tissue regenerations

Abstract: 'Heart' is the first organ to become fully functional during the development of an organism. However, it is the most complicated organ to regenerate when it is damaged. Humans cannot live without it, so tremendous efforts have been paid to recover the damaged heart. Recent studies indicate that mammalian cardiomyocytes can self-regenerate during normal aging as well as under pathological conditions. However, the source of such renewal remains still elusive. In this chapter, the use of cardiac stem cells to utilize knowledge that has been generated *in silico* is discussed. The text in this chapter has been modified from an original publication in InTech: http://www.intechopen.com/source/pdfs/18225/InTech-Are_we_there_yet_a_story_about_cardiac_stem_cells.pdf.

Keywords: adult stem cells, Cre-recombinase, heart, lineage tracing, reporter mice, tet-off, transgenic mice

6.1 Introduction

The heart is the first organ to become fully functional during the development of an organism, and many phrases (e.g. 'from the bottom of one's heart', 'get to the heart of') emphasize its importance. Heart dysfunction deteriorates the quality of one's life and ultimately leads to death. Therefore, tremendous efforts have been paid to recover damaged heart tissue. However, there is as yet no definitive cure for cardiovascular disease. The number of deaths related to cardiovascular diseases (CVD) emphasizes this point (http://www.who.int/cardiovascular_diseases/en/), as it is currently the number one cause of death worldwide and will remain so for years to come. Additionally, the causes of CVD are varied (environmental, genetic, life style, physiological, etc.; Kullo and Cooper, 2010), which makes finding cures extremely difficult.

Following the first study reporting heart as a post-mitotic organ, which concluded that hypertrophy is caused by enlargement rather than proliferation of cardiomyocytes (Karsner *et al.*, 1925), the view that heart can be regenerated, which persisted from 1850 to the first quarter of the 20th century, was overthrown (Carvalho and de Carvalho, 2010). Until the series of publications by Dr Piero Anversa and his colleagues from the 1980s, the view that heart (i.e. cardiomyocytes) cannot be renewed was firmly set in the field of Cardiology and Biology in general. Although the articles by Anversa *et al.* are often viewed as controversial (e.g. Anversa *et al.*, 2007), one of the most cited papers in Cardiovascular Medicine was published in 2009, firmly supporting Dr Anversa's view. This study was conducted by Dr Jonas Frisén and his colleagues (Bergmann *et al.*, 2009) by taking advantage of what proved to be a human disaster, namely the nuclear testing done until 1963, which labeled humans with carbon-14 (C14). This was a real-life pulse-and-chase experiment as C14 in the atmosphere was incorporated into plants and animals or diffused into the oceans and thus entered the food cycle. Using the C14 concentration in cardiomyocyte genomic DNA to calculate the age of cardiomyocytes, the authors estimated that human cardiomyocytes turnover at a rate of 1% per year at the age of 20, declining to 0.45% at the age of 75. According to their calculations, fewer than 50% of cardiomyocytes are renewed during a normal human life span. Dr Anversa's group has since reported a much higher rate of turnover. Using cancer patients who received infusion of the radiosensitizer iododeoxyuridine for therapy, which rapidly incorporated into cycling cells to allow for a pulse-and-chase experiment (similar to bromodeoxyuridine-labeling; DuFrain *et al.*, 1984), on average, the turnover of cardiomyocytes was 22% per year (20 and 13% for fibroblasts and endothelial cells, respectively) and their life span was found to be ~4.5 years, which corresponds to several cardiomyocyte renewals during a normal life span (Kajstura *et al.*, 2010b). To further support this, Dr Anversa's group published another report (Kajstura *et al.*, 2010a), which extended their previous findings that the aging human hearts of women are more adoptive to cardiomyocyte loss that those of men (Olivetti *et al.*, 1995). Based on their previous observation, they hypothesize that the gender difference in the aging-dependent adaptation of cardiomyocytes might be due to gender-specific differences in cardiomyocyte turnover. By counting cardiomyocytes and c-Kit+ cardiac stem cells (CSCs) undergoing apoptosis, cellular senescence and proliferation, the authors concluded that from 20 to 100 years of age, the entire myocyte compartment of women is replaced 15 times, whereas that of men was replaced 11 times. These numbers, which are rather high, are awaiting independent confirmation from other labs.

At first glance, the results of the above three reports in human might contradict findings in mice published in 2007. By utilizing transgenic mice to trace the fate of adult cardiomyocytes based on alpha-major histocompatibility complex-driven green fluorescent protein (GFP) expression, Hsieh and colleagues observed cardiomyocyte turnover by stem or precursor cells after myocardial infarction or pressure overload. However, during normal aging up to 1 year, equivalent to 34 years in human (Holaska, 2008), they did not record significant turnover of cardiomyocytes (Hsieh *et al.*, 2007). Labeling based on tamoxifen injection was not 100% (82.7 ± 1.7% in the above study), and there was a fluctuation of 2% of labeled cells, which might reflect the results of Bergmann *et al.* (2009) in human heart.

Although the above studies in human cardiomyocyte turnover are intriguing and provide impressive results, controversy remains. As human cardiomyocytes initiate cell cycle activity in response to mechanical stress (e.g. hypertrophy) without nuclear or cell division (karyo- and cyto-kinesis, respectively) (Adler and Friedburg, 1986; Soonpaa and Field, 1998; Wohlschlaeger *et al.*, 2010), this might increase the number of labeled cardiomyocytes, obscuring their calculations. However, it should be pointed out that unlike murine cardiomyocytes, which are >90% binucleated (Soonpaa *et al.*, 1996; Liu *et al.*, 2010), 74% of human cardiomyocytes are mononucleated, compared with 25.5% for bi-, 0.4% for tri- and 0.1% for tetranucleated cardiomyocytes (Olivetti *et al.*, 1996), which is the point addressed by Dr Anversa's group (Kajstura *et al.*, 2010a, b). This view of mononucleated human cardiomyocytes was recently challenged by Dr Frisén's group. According to their calculations, the majority of cardiomyocyte nuclei from the human adult left ventricles have more than two complete sets of chromosomes, 33.5, 55.8 and 10.7% being diploid, tetraploid and octaploid, respectively (Bergmann *et al.*, 2011). Therefore, the controversy continues. However, it should not prove too difficult to resolve this discrepancy in the future.

Given all the above findings, it is now reasonable to state that cardiomyocytes can be renewed during the lifetime of a higher organism. If this holds, then the interesting question that needs to be answered is the origin of such turnover, evidence for which was provided by Dr Anversa's group for c-Kit+ human CSCs (Kajstura *et al.*, 2010a). Their findings are intriguing, but to find a way to activate such CSCs, it is imperative to use a model organism, as we cannot experiment with humans. Therefore, in this chapter, the origin of cells in heart is first surveyed, and the current status of CSCs is investigated by paying particular attention to mouse as a model organism. I finish with a proposed model to study the source of cardiomyocyte turnover.

6.2 Cell types in heart and their lineages

Given its primary function as a pump, the heart is often viewed as mainly comprising muscle (i.e. cardiomyocytes). Mouse heart consists of ~56% cardiomyocytes, 27% fibroblasts, 10% vascular smooth muscle cells and 7% endothelial cells (Banerjee *et al.*, 2007). Interestingly, these cell type proportions vary within different rodent species. In the rat heart, cardiomyocytes make up only 26.4%, which is similar to that of human, and fibroblasts are the major cell type (62.6%) (Nag, 1980; Banerjee *et al.*, 2007). Based just on the composition of cell types in heart, rats should more reflect the situation of human heart (Rubart and Field, 2006). Indeed, more physiological experiments are done using rats than mice (Aitman *et al.*, 2008). However, until the recent development of zinc finger nucleases (Geurts *et al.*, 2009) and rat embryonic stem (ES) cells (Buehr *et al.*, 2008; Li *et al.*, 2008), the creation of transgenic rats was not possible (Tong *et al.*, 2010; Cui *et al.*, 2011), and thus mice tended to be used as a model organism. Given that the cell type composition differs greatly between human and mice, it might not be possible to translate all the findings in mice to the development of therapeutic methods in human, but the availability of various knockout and transgenic mice is still attractive for elucidation of mechanisms of CSCs. In the following subsections, the cell types in heart will be discussed in conjugation with CSCs.

6.2.1 Cardiomyocytes

Being the functional part of a heart that produces the driving force to pump blood around the body, cardiomyocytes are the priority concern in the field of Cardiovascular Regenerative Medicine. Despite the drawback set by the misconception of cardiomyocyte turnover (Karsner *et al.*, 1925), there have been many studies investigating the birth and growth of cardiomyocytes. Of these, a study conducted by Dr Loren Field's group is considered to be one of the most carefully done studies about the growth of postnatal cardiomyocytes (Soonpaa *et al.*, 1996). According to this study, cardiomyocyte DNA synthesis occurs in two distinct phases during murine development. The first phase is associated with cardiomoycyte proliferation, which occurs during fetal life. The second phase follows after the cession of reduplication of cardiomyocytes (the transition from the first phase takes place before postnatal day 10). The feature of this phase is binucleation of cardiomyocytes due to a round of genomic duplication and karyokinesis without cytokinesis.

There are speculated to be four sources of cardiomyocyte renewal (Parmacek and Epstein, 2009):

1. Adult cardiomyocytes re-entering the cell cycle and dividing
2. Bone-marrow-derived cardiac stem/progenitor cells
3. Cells derived from the embryonic epicardium
4. Cardiac stem/progenitor cells

Because a lack of sufficient numbers of cardiomyocytes can cause many forms of congenital and adult cardiovascular diseases, intensive research has been conducted to find a set of genes/proteins that might push adult cardiomyocytes into cell cycle to regenerate cardiomyocytes (Rubart and Field, 2006). However, it needs to be borne in mind that such a method is similar to what is happening in the progression of tumor growth, where cells divide uncontrollably. As primary cardiac tumors are rare (Devbhandari *et al.*, 2007), there must be an intrinsic, biological block that prevents division of cardiomyocytes soon after birth. Studies by several laboratories have shown that it is difficult to overcome this block and to achieve controlled proliferation of adult cardiomyocytes (Ebelt *et al.*, 2005, 2006, 2008a,b). There is still hope, however, to be able to utilize the remaining, low-level potential of cardiomyocytes to proliferate for therapeutic purposes.

The second speculated source of cardiomyocyte renewal is from bone-marrow-derived mesenchymal stem cells (BMSCs). The transdifferentiation of such cells has been much debated, and readers interested in this are referred to the articles by Phinney and Prockop (2007), Psaltis *et al.* (2008) and Alaiti *et al.* (2010). The current consensus in the field of Regenerative Medicine is that BMSCs are beneficial in treating patients with acute myocardial infarction and ischemic heart failure (Chugh *et al.*, 2009). However, transdifferentiation of transplanted or injected BMSCs (fresh or cultured) is not very likely to contribute to the success of these therapies; instead, BMSCs secrete growth factors and cytokines that might enhance survival of surviving cardiomyocytes and stimulate endogenous repair mechanisms via activation of resident CSCs and other stem cells (Wen *et al.*, 2010). Regarding the issue of transdifferentiation, an interesting study was conducted by Nern *et al.* (2009). In their study, the authors utilized a lineage-tracing system by targeting the hematopoietic-specific promoter *vav* to monitor cell fusion events under physiological conditions to challenge transplantation studies using BMSCs. In the case of heart, the authors found a single LacZ-positive (the reporter gene from the Rosa26 LacZ allele) cardiomyocyte from four hearts of non-irradiated healthy transgenic mice (vav-iCre/Rosa26 LacZ). Therefore, it is unlikely that hematopoietic cells contribute to renewal of cardiomyocytes. Of course, one could argue that such cells are different from BMSCs as they should be 'stem cells' of hematopietic cells; however, taken together with the clinical data, this indicates that BMSCs may not be the major player for cardiomyocyte renewal.

The third source of cardiomyocyte renewal is so-called 'epicardially derived mesenchymal cells (EPDCs)' (Morabito *et al.*, 2001). EPDCs are derived from a subpopulation of epicardial cells and are shown to differentiate into cardiac vessels, cardiomyocytes and connective tissue of heart (Limana *et al.*, 2011). Sometimes, they are considered as 'cardiac stem cells' (Wessels and Pérez-Pomares, 2004). The series of articles published by Dr Paul Riley's group indicate that EPDCs are dormant ('resting') in normal physiological settings (Smart, 2007; Smart and Riley, 2009). However, upon an appropriate stimulation by the actin monomer binding protein Thymosin β4, EPDCs were shown to have the capacity to proliferate, migrate and differentiate into vascular endothelial cells, smooth muscle cells and cardiac fibroblasts as well as cardiomyocytes (Smart *et al.*, 2010, 2011). Taking these data together, this leaves us with resident cardiac stem/progenitor cells as the most promising source of cardiomyocyte renewal.

6.2.2 Smooth muscle cells

Smooth muscle is an involuntary non-striated muscle that, unlike its cousins (cardiac and skeletal muscle cells), is highly plastic and can modulate between contractile and synthetic phenotypes in response to extracellular cues (Owens, 1995). In heart, vascular smooth muscle is a type of smooth muscle that contracts and regulates blood vessel tone, blood pressure and blood flow (Rzucidlo *et al.*, 2007), which are integral to the function of heart. Given its importance, heart dysfunctions lead to variety of diseases, including atherosclerosis, which results in over 55% of all deaths in Western countries (Owens *et al.*, 2004). Therefore, significant efforts have been spent to elucidate the pathological mechanisms in relation to vascular smooth muscle cells as well as develop a cure for such diseases. One attractive therapy is the use of adult vascular progenitor cells (Kumar and Caplice, 2010). As the major source of such progenitor cells is from vascular walls, which consist of endothelial cells, they will be discussed in the next subsection.

6.2.3 Endothelial cells

In heart, endothelial cells go hand in hand with vascular smooth muscle cells as they are the building blocks of arteries and veins, which are necessary for proper circulation of blood. Although there are no smooth muscle cells in capillaries, which are composed of endothelial cells, in arteries and veins, endothelial cells comprise the innermost layer called '(tunica) intima' whereas smooth muscle cells occupy the '(tunica) media', the middle layer of an artery or vein (Ergun *et al.*, 2011). Given their importance, endothelial cells have been intensively studied, and numerous excellent review articles have been published, which can be searched through PubMed. Therefore, I focus here only on 'vascular progenitor cells', also known as 'endothelial progenitor cells (EPCs)'.

Since the first report of EPCs by Asahara *et al.* (1997), much detailed research has been conducted to characterize this cell type. As in any other stem cell research studies, researchers mainly relied on surface markers to identify such cells, sort them and put them in culture to differentiate them *in vitro* to test their plasticity. Such studies revealed lab-to-lab, equipment-to-equipment and reagent-to-reagent (e.g. antibodies) differences, which led to discussions of contaminations by different cell types (e.g. platelets; Prokopi *et al.*, 2009). However, it should be noted that regardless of such controversies, the significance of EPCs in the diagnosis of cardiovascular diseases is undoubted (Boos *et al.*, 2006; Smadja *et al.*, 2009). Furthermore, the clinical potential of EPCs is relatively high, and a number of clinical trials have been conducted (Kumar and Caplice, 2010).

What is most confusing is that there is no defined marker nor any clear niche for EPCs identified (Psaltis *et al.*, 2010; Ergun *et al.*, 2011; Torsney and Xu, 2011). This led to the proposal that most EPCs are members of the hematopoietic lineage (Richardson and Yoder, 2011).

Given that heart is the place for blood circulation, it is probably the most likely organ to be utilizing such hematopoietic lineage cells, which are also classified as EPCs by others, for tissue homeostasis; this is by no means limited to cardiomyocytes. However, as pointed out previously, a defined lineage tracing using the hematopietic-specific promoter *vav* (Nern *et al.*, 2009) clearly indicates that a contribution of hematopoietic cells to the tissue homeostasis of heart is unlikely (Dr Stefan Momma, pers. comm.). Therefore, it is most likely that some EPCs are different from those of hematopoietic lineage cells. However, we need to keep in mind that both matured endothelial and smooth muscle cells are plastic, which allow them to dedifferentiate and to transdifferentiate themselves to other cell types (e.g. endothelial–mesenchymal transition) under pathological conditions (Zeisberg *et al.*, 2007; Speer *et al.*, 2009; Stintzing *et al.*, 2009). The above findings make it difficult to assess the contribution of such EPCs to cardiomyocyte renewal without a clearly defined lineage tracing study targeting a marker for EPCs, which is currently lacking.

6.2.4 Cardiac fibroblasts

Perhaps one of the most surprising recent findings was induced pluripotent stem cells (iPS cells) (Takahashi and Yamanaka, 2006). With the introduction of four transcription factors, *Oct-3/4* (*Pou5f1*), *Sox2*, *c-Myc* and *Klf4* (now called 'Yamanaka factors'), fibroblasts can be reprogrammed to be morphologically and biochemically similar to pluripotent stem cells. Recent studies indicate that the epigenetic status and gene expression profile of individual iPS isolates become similar to and indistinguishable from those of ES cells when iPS cells are passaged several times (Guenther *et al.*, 2010; Kim *et al.*, 2010; Polo *et al.*, 2010). Fibroblasts in heart, called 'cardiac fibroblasts', also show this phenomenon. Recently, Ieda *et al.* (2010) reprogrammed adult cardiac fibroblasts using three transcription factors, *Gata4*, *Mef2c* and *Tbx5*, which are important for cardiac development. Furthermore, Efe *et al.* (2011) succeeded in converting fibroblasts into beating atria-like cardiomyocytes in just 12 days without fully reprogramming them into iPS cells.

Although the reprogramming by defined transcription factors is clear, the above findings are confusing given that the origin of fibroblasts is currently unknown, primarily due to a lack of well-defined fibroblast-specific markers. Current consensus is that fibroblasts are derived from local epithelial and endothelial cells through the epithelial–mesenchymal transition (Neilson, 2010). Primarily due to this process, fibroblasts comprise a heterogeneous population of cells. Given that the efficiency of reprogramming of isolated fibroblasts is very low (0.5%; Jaenisch and Young, 2008) and that by inactivating p53 this proportion goes up to 10% (Hong *et al.*, 2009), there should be a population of fibroblasts that are more susceptible to reprogramming than others. This point has been raised by Dr Shinya Yamanaka's group, which termed them 'good' and 'bad' iPS cells based on teratoma-forming propensity (Miura *et al.*, 2009).

Intensive research has shown that it is possible to reprogram with only one factor in the case of neural stem cells by ectopic expression of OCT4 alone (Kim *et al.*, 2009a,b). This evidence questions what really defines mature cells; is it really correct to call fibroblasts as a mature and final form of cell types given that they are known to be plastic (Eyden, 2004)? This point becomes much clearer when one considers the morphological observation of primary BMSCs in culture, which are described to have 'fibroblast-like' morphology (Friedenstein *et al.*, 1976). The same morphological description is always used to describe adult stem cells in culture (Rios and Williams, 1990); also, CSCs are not exceptional (Messina *et al.*, 2004). By definition, CSCs should give rise to all cell lineages in heart, including cardiac fibroblasts. Given that no defined marker for fibroblasts has been found and given their plasticity in the case of iPS cells, we may need to consider fibroblasts as a transition to mature cell types from CSCs.

6.3 Current status of cardiac stem cells

Currently, there are six schools of thought regarding CSCs (Table 6.1). For obvious reasons, tremendous attention has been paid to characterize CSCs using various techniques. Later, an increasing number of review articles cited the same studies over and over again to firmly set the field for CSCs (Anversa *et al.*, 2006; Tateishi *et al.*, 2008; Ausoni and

Table 6.1 List of resident CSCs

Type of CSCs	Other markers	References
Sca1+ CSCs	CD34–, CD45–, FLK1–, c-kit+/–, GATA4+, NKX2–5+/–, MEF2C+	(Oh *et al.*, 2003; Matsuura *et al.*, 2004; Rosenblatt-Velin *et al.*, 2005; Wang *et al.*, 2006; Wu *et al.*, 2006; Tateishi *et al.*, 2007; Forte *et al.*, 2008)
c-kit+ CSCs	CD34–, CD45–, Sca1+, GATA4+, NKX2–5+, MEF2C+	(Beltrami *et al.*, 2003; Urbanek *et al.*, 2003; Dawn *et al.*, 2005; Linke *et al.*, 2005; Bearzi *et al.*, 2007; Tillmanns *et al.*, 2008; Miyamoto *et al.*, 2010)
Isl-1+ CSCs	CD31–, Sca1–, c-kit–, GATA4+, NKX2–5+	(Laugwitz *et al.*, 2005; Moretti *et al.*, 2006)
Side population (SP) cells	CD34+, CD45+, ABCG2+, Sca1+, c-kit+, NKX2–5–, GATA4–	(Martin *et al.*, 2004; Pfister *et al.*, 2005; Oyama *et al.*, 2007; Liang *et al.*, 2010)
Cardiospheres	CD29+, CD34+, CD90+, CD105+, FLK-1+, Sca1+, c-kit+	(Messina *et al.*, 2004; Smith *et al.*, 2007; Tateishi *et al.*, 2007; Andersen *et al.*, 2009; Cheng *et al.*, 2010; Davis *et al.*, 2010)
Cardiac mesangioblasts (EPCs)	CD31+, CD34+, CD44+, CD45–, Sca1+, c-kit+	(Galvez *et al.*, 2008, 2009; Barbuti *et al.*, 2010)

Sartore, 2009; Musunuru *et al.*, 2010; Di Nardo *et al.*, 2010; Bollini *et al.*, 2011). If one looks only at the number of types of CSCs, it seems that heart, once considered to be a post-mitotic organ, harbors the greatest number of adult stem cells. And some researchers are currently asking whether it is correct to call all CSCs 'true' adult stem cells with multipotency (Stamm *et al.*, 2009; Ellison *et al.*, 2010). Such a view can be understood if one considers the current status of the hierarchy of hematopoietic stem cells, which consists of multipotent adult stem cells and progenitor cells (sometimes classified as 'trans-amplifying cells') (Bryder *et al.*, 2006). In fact, Anversa *et al.* (2006) had already proposed such a hierarchy for CSCs.

As described in the previous sections, postnatal cardiomyocytes can be renewed throughout the life of a mammal. Given the evidence provided above, it is most likely that such turnover derives from

CSCs. Whether such CSCs originate from bone marrow or locations besides heart needs to be clarified, but the studies cited above suggest that at least those from hematopoietic lineage cells are unlikely (Nern *et al.*, 2009). Recent findings of circulating EPCs and 'very small embryonic-like stem cells (VSELs)' (Ratajczak *et al.*, 2009; Shin *et al.*, 2009, 2010) have stimulated debate regarding the origin of CSCs, but studies conducted using such 'circulating' ('non-resident') cells have been based only on *ex vivo* experiments: isolate such cells using antibodies against surface markers and inject them into another animal ('host'). In essence, this is the same procedure as with any other bone marrow transplantation experiments, which often results in the fusion events to cells of the host animal or uptake of markers [e.g. green fluorescent protein (GFP)] by the host cells, which is the very point that Nern *et al.* (2009) addressed. The fusion event is of concern whenever such studies are conducted (Wurmser and Gage, 2002). However, given the nature of postnatal cadiomyocytes (binucleation of cardiomyocytes due to a round of genomic duplication and karyokinesis without cytokinesis), we are likely to find cardiomyocytes with two nuclei, which is not the result of a so-called 'fusion' event. As no clear fusion event of a cardiomyocyte with another cell type in heart (e.g. endothelial cells) *in vivo* under normal physiological conditions has been shown, it seems pointless to argue whether such a fusion event is an artefact of the results obtained from such experiments in heart. Other studies have used a more direct injection into the heart after damage (e.g. myocardial infarction) (Chavakis *et al.*, 2010; Domian *et al.*, 2010; Maltais *et al.*, 2010; Siu *et al.*, 2010) and through the use of a sheet of cells (Matsuura *et al.*, 2009), but the intrinsic question that remains is what happens in natural settings *in vivo*. This point can be answered only when one considers a lineage tracing experiment targeting a proposed marker for CSCs.

6.3.1 Lineage tracing

As indicated above, the 'gold standard' in the field of stem cell research is lineage tracing by targeting a particular stem cell marker.

By definition, lineage tracing experiments trace the fates of stem cells using model organisms based on transgenic technology. One needs to differentiate between transient and permanent labeling. Transient labeling is based on knock-in of a reporter gene (e.g. GFP) into a gene of interest to make a fusion protein. This has been successfully applied to observe the contributions of c-kit+ cells to the revascularizing infarct regions of the myocardial infarcted heart by utilizing c-kit(BAC)-EGFP mice (Tallini *et al.*, 2009). These labeled cells in the infarct regions were not cardiomyocytes but endothelial and smooth muscle cells. As for labeling of c-kit+ cells under normal physiological conditions, the authors observed an increasing number of labeled cells in heart up to postnatal day 2. Thereafter, the number declined, and the labeled cells were rarely observed in adult heart. From these results, the authors concluded that 'c-kit expression in cardiomyocytes in the adult heart after injury does not identify cardiac myogenesis.' (Tallini *et al.*, 2009). This is due to the limitation of transient labeling, where once the labeled cells differentiate, they will turn the targeted adult stem cell marker gene off: this will result in no labeling.

In contrast, a permanent labeling strategy can overcome the problem of no labeling after differentiation. This can be done by utilizing the Cre/loxP system crossed to a reporter line (Smedley *et al.*, 2010). Some researchers use an inducible system [e.g. tamoxifen-inducible Mer-Cre-Mer system (Petrich *et al.*, 2003)] to give tighter control over labeling efficiencies. Alternatively, instead of double transgenic mice (Cre-reporter), triple transgenic mice based on the tet-Cre system can be utilized (Tang *et al.*, 2008).

6.3.2 Proposed study

6.3.2.1 Permanent labeling strategy

Here, we propose a permanent labeling strategy to lineage-trace the fates of adult stem cells. As shown in Figure 6.1, a tetracycline transactivator (tTA) is placed in the promoter region of the target gene to create a knock-in mouse. This system is called 'tet-off', in

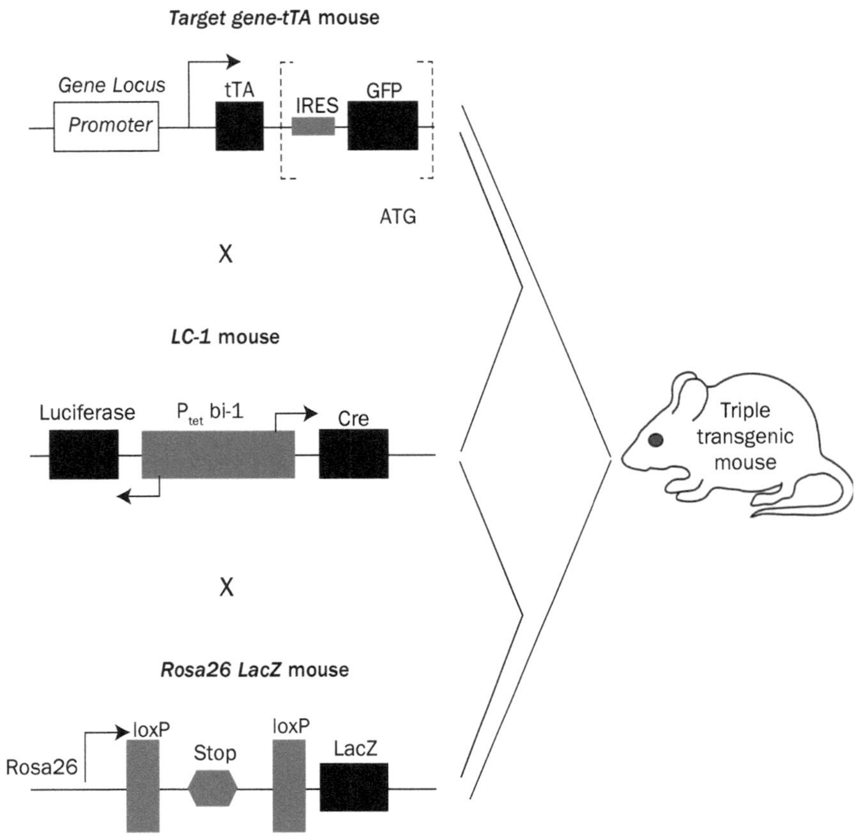

Figure 6.1 Triple transgenic mouse model

which tTA is functional until doxycycline (DOX) is added (Gossen and Bujard, 1992; Urlinger *et al.*, 2000).

The opposite system is called 'tet-on' (rtTA is used), which is claimed to be more tightly controlled (Stary *et al.*, 2010). However, when the tet-on system is used, DOX must be administered in the drinking water of mice, which can be very costly. In addition, DOX inhibits angiogenesis in mouse (Fainaru *et al.*, 2008; Cox *et al.*, 2010). In our lab, the DOX-treated murine hearts tend to be ~30% smaller than age-matched untreated hearts (data not shown).

Furthermore, DOX can have a positive effect on the attenuation of cardiac hypertrophy through inhibition of matrix metalloproteases (Errami *et al.*, 2008), which makes it difficult to challenge such mice to transverse aortic constriction (TAC) to observe the contribution of marked adult stem cells to the regeneration and remodeling of hypertrophied hearts. A recent study has shown that DOX can suppress doxorubicin-induced oxidative stress and cellular apoptosis in mouse heart (Lai *et al.*, 2010), which is another cardiovascular disease model commonly used in the field of Cardiovascular Medicine.

To monitor the expression of a target gene, one could place a reporter gene (e.g. GFP) through the use of an internal ribosomal entry site element (Attal *et al.*, 1999). However, depending on the construct, the expression of a reporter gene cannot be detected due to its low expression (Mizuguchi *et al.*, 2000). In addition, if a fluorescent reporter gene is used, this limits the number of fluorescent signals that can be used for further studies to identify the fates of once adult stem cells. Furthermore, there exists autofluorescence of heart, due to lipofuscin, which is a breakdown product of old red blood cells (Van de Lest *et al.*, 1995). These lipofuscin granules overlap with the fluorescent spectrum of the GFP signal, hampering interpretation of the experimental results (Laflamme and Murry, 2005). Therefore, the use of such reporter genes must be considered carefully.

When such a tTA mouse is created, it should be crossed with a mouse containing a tTA/rtTA responsive element (TRE), such as 'Ptet-1' (Baron and Bujard, 2000). Figure 6.1 shows mouse line 'LC-1' as an example, which ubiquitously expresses Cre-recombinase and the luciferase gene in all tissues in an adult mouse upon induction (Schönig *et al.*, 2002). As shown by Schönig *et al.* (2002), this mouse line was used in the triple transgenic mouse system based on tet-Cre crossed to a reporter line as we propose here. Alternatively, Tet-O-Cre transgenic mice, which create TRE-directed Cre-recombinase, can be used (Radomska *et al.*, 2002; Le *et al.*, 2008; Tang *et al.*, 2008; Hsu *et al.*, 2010). The offspring of this mating are termed 'tet-Cre mice'.

Upon successful creation of tet-Cre mice, these mice should be crossed to a reporter line that permanently labels the cells that express tTA when a target gene is expressed, which leads to creation of Cre-recombinase through the activity of TRE. This labeling must be genetically based so that when an adult stem cell that expressed a target gene differentiates into a mature cell (e.g. cardiomyocyte), this cell should be able to be traced. There are a variety of reporter lines currently available:

- **Rosa26 LacZ** (Soriano, 1999), which produced LacZ expression at the Rosa26 locus
- **Z/AP** (Lobe *et al.*, 1999), which expresses the LacZ gene before Cre-mediated excision, and upon Cre excision, the human placental alkaline phosphatase (AP) gene is expressed
- **Z/EG** (Novak *et al.*, 2000), like Z/AP mice, which express the LacZ gene, and upon Cre excision, enhanced GFP is expressed.

The beauty of this mouse system is that not only those cells (including adult stem cells) that currently express the target stem cell marker but also those cells that once expressed the target gene and turned it off (as should be the case when a cell is no longer a stem cell; in other words, if this does not hold, then the target marker gene is not a stem cell marker) can be labeled. To maximize this feature, one could serially cross-section the heart of a triple transgenic mouse and stain for the marker protein ('LacZ' in Figure 6.1) together with tissue markers. In the case of heart, as described earlier in this chapter, adult murine myocardium is composed of cardiomyocytes, fibroblasts, and endothelial and vascular smooth muscle cells (Banerjee *et al.*, 2006, 2007). If what has been proposed and validated in human is true, then the number of cardiomyocytes arising from various sources (e.g. bone marrow, resident CSCs) (Parmacek and Epstein, 2009) should increase as the animal ages. With our proposed mouse system, it is possible to validate if this holds true in mouse by simply counting the number of labeled cells using cardiomyocyte markers. To make this more statistically sound, we propose the following antibodies and fluorescently labeled substrates against tissue-specific proteins, and in the order shown here: BS1 isolectin B4 [known as an endothelial marker (Ziegelhoeffer *et al.*, 2004), but our preliminary results show that it also recognizes fibroblasts (data not shown)], dystrophin (to mark the cell membrane of cardiomyocytes to allow for easier identification), LacZ staining (Oustanina *et al.*, 2004), alpha smooth muscle actin (for smooth muscle cells) and wheat germ agglutinin (WGA) (a type of 'lectin') [it can bind oligosaccharides containing terminal *N*-acetylglucosamine or chitobiose, compounds that are common to many serum and membrane glycoproteins (Nagata and Burger, 1974)]. Given that WGA recognizes cell membranes of all cell types in heart, we could use the number of WGA-positive, labeled cells as a denominator and consider the number of tissue-type specifically labeled cells as a numerator to obtain a percentage of each tissue type contribution. Using this calculation scheme with mice of different ages, it is possible to obtain the percentage distribution of labeled cells, especially those of cardiomoycytes. To make this calculation statistically more valid, one could measure the size of each section of a heart to derive the number of labeled cells per square millimeter.

6.3.2.2 New and exciting reporter lines

A recent study indicates that the methylation of each transgenic promoter mentioned above differs greatly, which results in the inhibition of Cre-mediated recombination of the locus, leading to differences in labeling efficiencies (Long and Rossi, 2009). Therefore, not one but at least two or more reporter lines should be used to determine the efficiency of the labeling system being used. For the reporter line, there is a line termed '*Brainbow*', which allows for the production of up to 166 colors by altering sites for lox variants so that upon Cre excision, a random splicing event occurs to alter the combinations of fluorescent proteins (OFP, M-RFP, M-YFP and M-CFP) (Livet *et al.*, 2007). The original *Brainbow* mice are based on the *Thy1* promoter, which restricts its use to certain tissues (e.g. neurons). Recently, Dr Hans Clevers' group have modified this *Brainbow* construct by replacing the *Thy1* promoter with the Rosa26 locus, which allows for ubiquitous expression across all tissues, and named this line as '*R26R-Confetti*' (Snippert *et al.*, 2010). By using

R26R-Confetti mice as a reporter line, in principle, one could label a single adult stem cell that expresses a marker of interest with a unique color, which is different from others. When this cell proliferates and makes daughter cells, which will carry the same combination of fluorescent proteins and expresses the same color as the parent cell, one could identify the fates of this single population of adult stem cells. When this strategy is employed in heart, it is possible to identify which population of adult stem cells gives rise to all three lineages of heart [cardiomyocytes, endothelial cells and smooth muscle cells (Moretti *et al.*, 2006)]. Given that there is no ultimate single CSC marker that marks 'true' CSCs, by identifying such a population of cells, one could isolate such cells based on their fluorescent signal and profile them using transcriptomics and/or proteomics approaches, which should allow for the identification of further 'unique' surface markers (Uchida *et al.*, 2011). In line with the aims of this book, I propose to utilize *in silico* approaches to data-mine the transcriptomics and/or proteomics datasets generated from such populations of CSCs. As listed in Chapter 2, there are many useful databases in this respect, such as the Human Protein Atlas (http://www.proteinatlas.org//index.php) and StemBase (http://www.stembase.ca). If everything works as proposed above, one could generate another tet line based on such a marker gene to check the contribution thereof to further elucidate the mechanisms by which CSCs maintain their stemness and their differentiation pathways.

6.3.2.3 Moving from observation study to functional study

Alternatively, having such a mouse system, one could challenge these mice through various mechanical stresses [e.g. transverse aortic constriction (TAC), myocardial infarction] as well as injection of different drugs and small molecules to activate the labeled CSCs to differentiate into the cell type of one's choice (most likely, cardiomyocytes in our case). For this, we recently introduced a model to induce right ventricular hypertrophy through pulmonary artery clipping (PAC), which avoids detrimental right ventricular pressure overload, and thus allows long-term survival of operated mice to recapitulate pathognomonic changes of patients with congenital heart disease and pulmonary arterial hypertension (Kreymborg *et al.*, 2010). Given that the origins of the right ventricle and outflow tract are from the secondary heart field (Waldo *et al.*, 2001; Verzi *et al.*, 2005), our PAC model should primarily affect the regions of the secondary heart field so that the behavior of the descendants of once Isl1-positive CSCs (Cai *et al.*, 2003; Laugwitz *et al.*, 2005) can be monitored. Furthermore, by applying DOX before or after, and even until, damage, the labeling can be suppressed to monitor the contribution of labeled cells to delineate the effects of such damage to the labeled cells or labeling itself.

Once the fates of labeled cells are determined, the next question is the mechanism by which these cells differentiate into certain cell types. Recently, Dr Loren Field's group reported that unlike those isolated from neonatal hearts, c-kit+ CSCs in mouse adult hearts failed to undergo cardiomyogenic differentiation when co-cultured with fetal cardiomyocytes or when transplanted into normal or infarcted mouse adult hearts (Zaruba, 2010). This study clearly indicates the difficulty of performing an *in vitro* study with isolated CSCs. However, when the above-mentioned triple transgenic mouse system is used, in principle, one could perform an *in vivo* mechanistic study by using Cre-recombinase. Through the activity of TRE, Cre-recombinase will be produced, which in the above example allows for the labeling of cells through a reporter line. If the above triple transgenic mice were crossed with a conditional knockout mouse to give quadruple transgenic mice with a homogyzous allele for a gene to be knocked out, then one could analyze the fate of CSCs under the null condition of a target gene that is critical for the differentiation into a functional cell type (e.g. cardiomyocytes).

For example, one could cross the proposed triple transgenic mice with a *Gata4*-loxP mouse with one loxP site upstream of exon 3 and another loxP site downstream of exon 5, which encode both zinc finger DNA-binding domains and the nuclear localization signal that are essential for the function of GATA4 (Watt *et al.*, 2004) for two generations to generate Target gene-tTA // LC-1 Cre // Rosa26 LacZ // c*Gata4* (–/–) quadruple transgenic mice. According to this model, when a cell expresses a target gene (adult stem cell marker), it will produce tTA and activate TRE to produce Cre-recombinase. This Cre-recombinase will do two things: one is to label the cell through a reporter allele (LacZ); and the other is to delete exons 3–5 of the *Gata4* gene to disrupt the function of GATA4 in this cell. If GATA4 is indeed important for the differentiation of labeled CSCs into a cardiomyocyte, then one should not see labeled cardiomyocytes in an adult heart. One could also use DOX to delineate the embryonic and adult contributions of such labeled cells to the tissue homeostasis of heart through CSCs.

By altering a gene to be knocked out, a classical Developmental Biological study of gene knockout in conjugation with adult stem cell research can be conducted. For this, one could explore international large-scale mouse knockout programs (Friedel *et al.*, 2007; Guan *et al.*, 2010). Of these, the European Conditional Mouse Mutagenesis Program (EUCOMM) (www.eucomm.org/) and Knockout Mouse Project (KOMP) (www.komp.org/) would be useful for our purposes as both projects aim to make conditional alleles together with the LacZ gene as a reporter to target most mouse genes. For a specific example, please see section 5.4.3. A word of caution: if one chooses to use a mouse line from such a project, a reporter line must be changed from Rosa26 LacZ to another without LacZ to have a reporter gene different from the one used for a conditional allele.

6.4. Conclusion

Recent findings have overthrown the misconception that heart is a post-mitotic organ; instead, it is now firmly confirmed that heart is a dynamically replacing organ. As discussed in this chapter, heart is composed of various cell types, and to understand the tissue homeostasis of heart, one needs to dissect the results from such perspectives as well. Although the ultimate aim of Cardiovascular Regenerative Medicine is to rejuvinate the ailing heart by replacing damaged cardiomyocytes with new ones, it is necessary to consider the steps by which this could be achieved, by first identifying resident CSCs in heart. To do this, it is of utmost importance to perform lineage tracing experiments to clearly identify the contributions of CSCs to the maintenance of heart, including cardiomyocytes. One such method of lineage tracing is proposed in this chapter, and I hope that readers will perform such experiments to confirm or dispute the *in vitro* findings regarding CSCs.

Keeping in line with the aims of this book, once such a lineage tracing mouse system is established, one could utilize the *in silico* approaches proposed here to further dissect CSCs to answer long-standing questions in the field of Stem Cell Biology, such as: Where do adult stem cells come from? Why are they still dormant? How could we activate such cells into a cell type of interest? As heart is not the only organ with adult stem cells, such a lineage tracing mouse system should be useful to study adult stem cells in other body parts (brain, lung, mammary gland, etc.). This system will be a perfect model to evaluate the knowledge generated from the '1,000 Genes Project' in the context of adult stem cells.

References

Adler CP, Friedburg H. (1986) Myocardial DNA content, ploidy level and cell number in geriatric hearts: post-mortem examinations of human myocardium in old age. *J Mol Cell Cardiol* 18(1): 39–53.

Aitman TJ, Critser JK, Cuppen E, *et al.* (2008) Progress and prospects in rat genetics: a community view. *Nat Genet* 40(5): 516–22.

Alaiti MA, Ishikawa M, Costa MA. (2010) Bone marrow and circulating stem/progenitor cells for regenerative cardiovascular therapy. *Transl Res* 156(3): 112–29.

Andersen DC, Andersen P, Schneider M, Jensen HB, Sheikh SP. (2009) Murine "cardiospheres" are not a source of stem cells with cardiomyogenic potential. *Stem Cells* 27(7): 1571–81.

Anversa P, Kajstura J, Leri A, Bolli R. (2006) Life and death of cardiac stem cells: a paradigm shift in cardiac biology. *Circulation* 113(11): 1451–63.

Anversa P, Leri A, Rota M, *et al.* (2007) Concise review: stem cells, myocardial regeneration, and methodological artifacts. *Stem Cells* 25(3): 589–601.

Asahara T, Murohara T, Sullivan A, *et al.* (1997) Isolation of putative progenitor endothelial cells for angiogenesis. *Science* 275(5302): 964–7.

Attal J, Theron MC, Puissant C, Houdebine LM. (1999) Effect of intercistronic length on internal ribosome entry site (IRES) efficiency in bicistronic mRNA. *Gene Expr* 8(5–6): 299–309.

Ausoni S, Sartore S. (2009) The cardiovascular unit as a dynamic player in disease and regeneration. *Trends Mol Med* 15(12): 543–52.

Banerjee I, Yekkala K, Borg TK, Baudino TA. (2006) Dynamic interactions between myocytes, fibroblasts, and extracellular matrix. *Ann NY Acad Sci* 1080: 76–84.

Banerjee I, Fuseler JW, Price RL, Borg TK, Baudino TA. (2007) Determination of cell types and numbers during cardiac development in the neonatal and adult rat and mouse. *Am J Physiol Heart Circ Physiol* 293(3): H1883–91.

Barbuti A, Galvez BG, Crespi A, *et al.* (2010) Mesoangioblasts from ventricular vessels can differentiate in vitro into cardiac myocytes with sinoatrial-like properties. *J Mol Cell Cardiol* 48(2): 415–23.

Baron U, Bujard H. (2000) Tet repressor based systems for regulated gene expression in eukaryotic cells: principles and advances. *Methods Enzymol* 327: 401–21.

Bearzi C, Rota M, Hosoda T, *et al.* (2007) Human cardiac stem cells. *Proc Natl Acad Sci USA* 104(35): 14068–73.

Beltrami AP, Barlucchi L, Torella D, *et al.* (2003) Adult cardiac stem cells are multipotent and support myocardial regeneration. *Cell* 114(6): 763–76.

Bergmann O, Bhardwaj RD, Bernard S, *et al.* (2009) Evidence for cardiomyocyte renewal in humans. *Science* 324(5923): 98–102.

Bergmann O, Zdunek S, Alkass K, Druid H, Bernard S, Frisén J. (2011) Identification of cardiomyocyte nuclei and assessment of ploidy for the analysis of cell turnover. *Exp Cell Res* 317(2): 188–94.

Bollini S, Smart N, Riley PR. (2011) Resident cardiac progenitor cells: at the heart of regeneration. *J Mol Cell Cardiol* 50(2): 296–303.

Boos CJ, Lip GY, Blann AD. (2006) Circulating endothelial cells in cardiovascular disease. *J Am Coll Cardiol* 48(8): 1538–47.

Bryder D, Rossi DJ, Weissman IL. (2006) Hematopoietic stem cells: the paradigmatic tissue-specific stem cell. *Am J Pathol* 169(2): 338–46.

Buehr M, Meek S, Blair K, *et al.* (2008) Capture of authentic embryonic stem cells from rat blastocysts. *Cell* 135(7): 1287–98.

Cai CL, Liang X, Shi Y, *et al.* (2003) Isl1 identifies a cardiac progenitor population that proliferates prior to differentiation and contributes a majority of cells to the heart. *Dev Cell* 5(6): 877–89.

Carvalho AB, de Carvalho AC. (2010) Heart regeneration: past, present and future. *World J Cardiol* 2(5): 107–11.

Chavakis E, Koyanagi M, Dimmeler S. (2010) Enhancing the outcome of cell therapy for cardiac repair: progress from bench to bedside and back. *Circulation* 121(2): 325–35.

Cheng K, Li TS, Malliaras K, Davis DR, Zhang Y, Marbán E. (2010) Magnetic targeting enhances engraftment and functional benefit of iron-labeled cardiosphere-derived cells in myocardial infarction. *Circ Res* 106(10): 1570–81.

Chugh AR, Zuba-Surma EK, Dawn B. (2009) Bone marrow-derived mesenchymal stems cells and cardiac repair. *Minerva Cardioangiol* 57(2): 185–202.

Cox CA, Amaral J, Salloum R, *et al.* (2010) Doxycycline's effect on ocular angiogenesis: an in vivo analysis. *Ophthalmology* 117(9): 1782–91.

Cui X, Ji D, Fisher DA, Wu Y, Briner DM, Weinstein EJ. (2011) Targeted integration in rat and mouse embryos with zinc-finger nucleases. *Nat Biotechnol* 29(1): 64–7.

Davis DR, Kizana E, Terrovitis J, *et al.* (2010) Isolation and expansion of functionally-competent cardiac progenitor cells directly from heart biopsies. *J Mol Cell Cardiol* 49(2): 312–21.

Dawn B, Stein AB, Urbanek K, *et al.* (2005) Cardiac stem cells delivered intravascularly traverse the vessel barrier, regenerate infarcted myocardium, and improve cardiac function. *Proc Natl Acad Sci USA* 102(10): 3766–71.

Devbhandari MP, Meraj S, Jones MT, Kadir I, Bridgewater B. (2007) Primary cardiac sarcoma: reports of two cases and a review of current literature. *J Cardiothorac Surg* 2: 34.

Di Nardo P, Forte G, Ahluwalia A, Minieri M. (2010) Cardiac progenitor cells: potency and control. *J Cell Physiol* 224(3): 590–600.

Domian IJ, Buikema JW, de Boer RA, van der Meer P. (2010) Stem cells in heart failure. *Eur J Heart Fail* 12(7): 642–4.

DuFrain RJ, McFee AF, Linkous S, Jennings CJ, Lowe KW. (1984) In vivo SCE analysis using bromodeoxyuridine, iododeoxyuridine, and chlorodeoxyuridine. *Mutat Res* 139(2): 57–60.

Ebelt H, Hufnagel N, Neuhaus P, *et al.* (2005) Divergent siblings: E2F2 and E2F4 but not E2F1 and E2F3 induce DNA synthesis in cardiomyocytes without activation of apoptosis. *Circ Res* 96(5): 509–17.

Ebelt H, Liu Z, Müller-Werdan U, Werdan K, Braun T. (2006) Making omelets without breaking eggs: E2F-mediated induction of cardiomyoycte cell proliferation without stimulation of apoptosis. *Cell Cycle* 5(21): 2436–9.

Ebelt H, Zhang Y, Kampke A, *et al.* (2008a) E2F2 expression induces proliferation of terminally differentiated cardiomyocytes in vivo. *Cardiovasc Res* 80(2): 219–26.

Ebelt H, Zhang Y, Köhler K, *et al.* (2008b) Directed expression of dominant-negative p73 enables proliferation of cardiomyocytes in mice. *J Mol Cell Cardiol* 45(3): 411–9.

Efe JA, Hilcove S, Kim J, *et al.* (2011) Conversion of mouse fibroblasts into cardiomyocytes using a direct reprogramming strategy. *Nat Cell Biol* 13(3): 215–22.

Ellison GM, Galuppo V, Vicinanza C, *et al.* (2010) Cardiac stem and progenitor cell identification: different markers for the same cell? *Front Biosci (Schol Ed)* 2: 641–52.

Ergün S, Tilki D, Klein D. (2011) Vascular wall as a reservoir for different types of stem and progenitor cells. *Antioxid Redox Signal* 2011, 7 January [Epub ahead of print]

Errami M, Galindo CL, Tassa AT, Dimaio JM, Hill JA, Garner HR. (2008) Doxycycline attenuates isoproterenol- and transverse aortic banding-induced cardiac hypertrophy in mice. *J Pharmacol Exp Ther* 324(3): 1196–203.

Eyden B. (2004) Fibroblast phenotype plasticity: relevance for understanding heterogeneity in "fibroblastic" tumors. *Ultrastruct Pathol* 28(5–6): 307–19.

Fainaru O, Adini I, Benny O, *et al.* (2008) Doxycycline induces membrane expression of VE-cadherin on endothelial cells and prevents vascular hyperpermeability. *FASEB J* 22(10): 3728–35.

Forte G, Carotenuto F, Pagliari F, *et al.* (2008) Criticality of the biological and physical stimuli array inducing resident cardiac stem cell determination. *Stem Cells* 26(8): 2093–103.

Friedel RH, Seisenberger C, Kaloff C, Wurst W. (2007) EUCOMM – the European conditional mouse mutagenesis program. *Brief Funct Genomic Proteomic* 6(3): 180–5.

Friedenstein AJ, Gorskaja JF, Kulagina NN. (1976) Fibroblast precursors in normal and irradiated mouse hematopoietic organs. *Exp Hematol* 4(5): 267–74.

Galvez BG, Sampaolesi M, Barbuti A, *et al.* (2008) Cardiac mesoangioblasts are committed, self-renewable progenitors, associated with small vessels of juvenile mouse ventricle. *Cell Death Differ* 15(9): 1417–28.

Gálvez BG, Covarello D, Tolorenzi R, *et al.* (2009) Human cardiac mesoangioblasts isolated from hypertrophic cardiomyopathies are greatly reduced in proliferation and differentiation potency. *Cardiovasc Res* 83(4): 707–16.

Geurts AM, Cost GJ, Freyvert Y, *et al.* (2009) Knockout rats via embryo microinjection of zinc-finger nucleases. *Science* 325(5939): 433.

Gossen M, Bujard H. (1992) Tight control of gene expression in mammalian cells by tetracycline-responsive promoters. *Proc Natl Acad Sci USA* 89(12): 5547–51.

Guan C, Ye C, Yang X, Gao J. (2010) A review of current large-scale mouse knockout efforts. *Genesis* 48(2): 73–85.

Guenther MG, Frampton GM, Soldner F, *et al.* (2010) Chromatin structure and gene expression programs of human embryonic and induced pluripotent stem cells. *Cell Stem Cell* 7(2): 249–57.

Holaska JM. (2008) Emerin and the nuclear lamina in muscle and cardiac disease. *Circ Res* 103(1): 16–23.

Hong H, Takahashi K, Ichisaka T, *et al.* (2009) Suppression of induced pluripotent stem cell generation by the p53-p21 pathway. *Nature* 460(7259): 1132–5.

Hsieh PC, Segers VF, Davis ME, *et al.* (2007) Evidence from a genetic fate-mapping study that stem cells refresh adult mammalian cardiomyocytes after injury. *Nat Med* 13(8): 970–4.

Hsu W, Mirando AJ, Yu HM. (2010) Manipulating gene activity in Wnt1-expressing precursors of neural epithelial and neural crest cells. *Dev Dyn* 239(1): 338–45.

Ieda M, Fu JD, Delgado-Olguin P, *et al.* (2010) Direct reprogramming of fibroblasts into functional cardiomyocytes by defined factors. *Cell* 142(3): 375–86.

Jaenisch R, Young R. (2008) Stem cells, the molecular circuitry of pluripotency and nuclear reprogramming. *Cell* 132(4): 567–82.

Kajstura J, Gurusamy N, Ogórek B, *et al.* (2010a) Myocyte turnover in the aging human heart. *Circ Res* 2010, 18 November [Epub ahead of print]

Kajstura J, Urbanek K, Perl S, *et al.* (2010b) Cardiomyogenesis in the adult human heart. *Circ Res* 107(2): 305–15.

Karsner HT, Saphir O, Todd TW. (1925) The state of the cardiac muscle in hypertrophy and atrophy. *Am J Pathol* 1(4): 351–372.1.

Kim JB, Greber B, Araúzo-Bravo MJ, *et al.* (2009a) Direct reprogramming of human neural stem cells by OCT4. *Nature* 461(7264): 649–3.

Kim JB, Sebastiano V, Wu G, *et al.* (2009b) Oct4-induced pluripotency in adult neural stem cells. *Cell* 136(3): 411–9.

Kim K, Doi A, Wen B, *et al.* (2010) Epigenetic memory in induced pluripotent stem cells. *Nature* 467(7313): 285–90.

Kreymborg K, Uchida S, Gellert P, *et al.* (2010) Identification of right heart-enriched genes in a murine model of chronic outflow tract obstruction. *J Mol Cell Cardiol* 49(4): 598–605.

Kullo IJ, Cooper LT. (2010) Early identification of cardiovascular risk using genomics and proteomics. *Nat Rev Cardiol* 7(6): 309–17.

Kumar AH, Caplice NM. (2010) Clinical potential of adult vascular progenitor cells. *Arterioscler Thromb Vasc Biol* 30(6): 1080–7.

Laflamme MA, Murry CE. (2005) Regenerating the heart. *Nat Biotechnol* 23(7): 845–56.

Lai HC, Yeh YC, Ting CT, *et al.* (2010) Doxycycline suppresses doxorubicin-induced oxidative stress and cellular apoptosis in mouse hearts. *Eur J Pharmacol* 644(1–3): 176–87.

Laugwitz KL, Moretti A, Lam J, *et al.* (2005) Postnatal isl1+ cardioblasts enter fully differentiated cardiomyocyte lineages. *Nature* 433(7026): 647–53.

Le YZ, Zheng W, Rao PC, *et al.* (2008) Inducible expression of cre recombinase in the retinal pigmented epithelium. *Invest Ophthalmol Vis Sci* 49(3): 1248–53.

Li P, Tong C, Mehrian-Shai R, *et al.* (2008) Germline competent embryonic stem cells derived from rat blastocysts. *Cell* 135(7): 1299–310.

Liang SX, Tan TY, Gaudry L, Chong B. (2010) Differentiation and migration of Sca1+/CD31– cardiac side population cells in a murine myocardial ischemic model. *Int J Cardiol* 138(1): 40–9.

Limana F, Capogrossi MC, Germani A. (2011) The epicardium in cardiac repair: from the stem cell view. *Pharmacol Ther* 129(1): 82–96.

Linke A, Müller P, Nurzynska D, *et al.* (2005) Stem cells in the dog heart are self-renewing, clonogenic, and multipotent and regenerate infarcted myocardium, improving cardiac function. *Proc Natl Acad Sci USA* 102(25): 8966–71.

Liu Z, Yue S, Chen X, Kubin T, Braun T. (2010) Regulation of cardiomyocyte polyploidy and multinucleation by CyclinG1. *Circ Res* 106(9): 1498–506.

Livet J, Weissman TA, Kang H, *et al.* (2007) Transgenic strategies for combinatorial expression of fluorescent proteins in the nervous system. *Nature* 450(7166): 56–62.

Lobe CG, Koop KE, Kreppner W, Lomeli H, Gertsenstein M, Nagy A. (1999) Z/AP, a double reporter for cre-mediated recombination. *Dev Biol* 208(2): 281–92.

Long MA, Rossi FM. (2009) Silencing inhibits Cre-mediated recombination of the Z/AP and Z/EG reporters in adult cells. *PLoS One* 4(5): e5435.

Maltais S, Tremblay JP, Perrault LP, Ly HQ. (2010) The paracrine effect: pivotal mechanism in cell-based cardiac repair. *J Cardiovasc Transl Res* 3(6): 652–62.

Martin CM, Meeson AP, Robertson SM, *et al.* (2004) Persistent expression of the ATP-binding cassette transporter, Abcg2, identifies cardiac SP cells in the developing and adult heart. *Dev Biol* 265(1): 262–75.

Matsuura K, Nagai T, Nishigaki N, *et al.* (2004) Adult cardiac Sca-1-positive cells differentiate into beating cardiomyocytes. *J Biol Chem* 279(12): 11384–91.

Matsuura K, Honda A, Nagai T, *et al.* (2009) Transplantation of cardiac progenitor cells ameliorates cardiac dysfunction after myocardial infarction in mice. *J Clin Invest* 119(8): 2204–17.

Messina E, De Angelis L, Frati G, *et al.* (2004) Isolation and expansion of adult cardiac stem cells from human and murine heart. *Circ Res* 95(9): 911–21.

Miura K, Okada Y, Aoi T, *et al.* (2009) Variation in the safety of induced pluripotent stem cell lines. *Nat Biotechnol* 27(8): 743–5.

Miyamoto S, Kawaguchi N, Ellison GM, Matsuoka R, Shin'oka T, Kurosawa H. (2010) Characterization of long-term cultured c-kit+ cardiac stem cells derived from adult rat hearts. *Stem Cells Dev* 19(1): 105–16.

Mizuguchi H, Xu Z, Ishii-Watabe A, Uchida E, Hayakawa T. (2000) IRES-dependent second gene expression is significantly lower than cap-dependent first gene expression in a bicistronic vector. *Mol Ther* 1(4): 376–82.

Morabito CJ, Dettman RW, Kattan J, Collier JM, Bristow J. (2001) Positive and negative regulation of epicardial-mesenchymal transformation during avian heart development. *Dev Biol* 234(1): 204–15.

Moretti A, Caron L, Nakano A, *et al.* (2006) Multipotent embryonic isl1+ progenitor cells lead to cardiac, smooth muscle, and endothelial cell diversification. *Cell* 127(6): 1151–65.

Musunuru K, Domian IJ, Chien KR. (2010) Stem cell models of cardiac development and disease. *Annu Rev Cell Dev Biol* 26: 667–87.

Nag AC. (1980) Study of non-muscle cells of the adult mammalian heart: a fine structural analysis and distribution. *Cytobios* 28(109): 41–61.

Nagata Y, Burger MM. (1974) Wheat germ agglutinin. Molecular characteristics and specificity for sugar binding. *J Biol Chem* 249(10): 3116–22.

Neilson EG. (2010) The Jeremiah Metzger lecture. The origin of fibroblasts and the terminality of epithelial differentiation. *Trans Am Clin Climatol Assoc* 121: 240–50.

Nern C, Wolff I, Macas J, *et al.* (2009) Fusion of hematopoietic cells with Purkinje neurons does not lead to stable heterokaryon formation under noninvasive conditions. *J Neurosci* 29(12): 3799–807.

Novak A, Guo C, Yang W, Nagy A, Lobe CG. (2000) Z/EG, a double reporter mouse line that expresses enhanced green fluorescent protein upon Cre-mediated excision. *Genesis* 28(3–4): 147–55.

Oh H, Bradfute SB, Gallardo TD, *et al.* (2003) Cardiac progenitor cells from adult myocardium: homing, differentiation, and fusion after infarction. *Proc Natl Acad Sci USA* 100(21): 12313–8.

Olivetti G, Cigola E, Maestri R, *et al.* (1996) Aging, cardiac hypertrophy and ischemic cardiomyopathy do not affect the proportion of mononucleated and multinucleated myocytes in the human heart. *J Mol Cell Cardiol* 28(7): 1463–77.

Olivetti G, Giordano G, Corradi D, *et al.* (1995) Gender differences and aging: effects on the human heart. *J Am Coll Cardiol* 26(4): 1068–79.

Oustanina S, Hause G, Braun T. (2004) Pax7 directs postnatal renewal and propagation of myogenic satellite cells but not their specification. *EMBO J* 23(16): 3430–9.

Owens GK. (1995) Regulation of differentiation of vascular smooth muscle cells. *Physiol Rev* 75(3): 487–517.

Owens GK, Kumar MS, Wamhoff BR. (2004) Molecular regulation of vascular smooth muscle cell differentiation in development and disease. *Physiol Rev* 84(3): 767–801.

Oyama T, Nagai T, Wada H, *et al.* (2007) Cardiac side population cells have a potential to migrate and differentiate into cardiomyocytes in vitro and in vivo. *J Cell Biol* 176(3): 329–41.

Parmacek MS, Epstein JA. (2009) Cardiomyocyte renewal. *N Engl J Med* 361(1): 86–8.

Petrich BG, Molkentin JD, Wang Y. (2003) Temporal activation of c-Jun N-terminal kinase in adult transgenic heart via cre-loxP-mediated DNA recombination. *FASEB J* 17(6): 749–51.

Pfister O, Mouquet F, Jain M, *et al.* (2005) CD31– but not CD31+ cardiac side population cells exhibit functional cardiomyogenic differentiation. *Circ Res* 97(1): 52–61.

Phinney DG, Prockop DJ. (2007) Concise review: mesenchymal stem/multipotent stromal cells: the state of transdifferentiation and modes of tissue repair–current views. *Stem Cells* 25(11): 2896–902.

Polo JM, Liu S, Figueroa ME, *et al.* (2010) Cell type of origin influences the molecular and functional properties of mouse induced pluripotent stem cells. *Nat Biotechnol* 28(8): 848–55.

Prokopi M, Pula G, Mayr U, *et al.* (2009) Proteomic analysis reveals presence of platelet microparticles in endothelial progenitor cell cultures. *Blood* 114(3): 723–32.

Psaltis PJ, Zannettino AC, Worthley SG, Gronthos S. (2008) Concise review: mesenchymal stromal cells: potential for cardiovascular repair. *Stem Cells* 26(9): 2201–10.

Psaltis PJ, Harbuzariu A, Delacroix S, Holroyd EW, Simari RD. (2010) Resident vascular progenitor cells-diverse origins, phenotype, and function. *J Cardiovasc Transl Res* 2010, 30 November [Epub ahead of print]

Radomska HS, Gonzalez DA, Okuno Y, *et al.* (2002) Transgenic targeting with regulatory elements of the human CD34 gene. *Blood* 100(13): 4410–9.

Ratajczak MZ, Kucia M, Ratajczak J, Zuba-Surma EK. (2009) A multi-instrumental approach to identify and purify very small embryonic like stem cells (VSELs) from adult tissues. *Micron* 40(3): 386–93.

Richardson MR, Yoder MC. (2011) Endothelial progenitor cells: Quo Vadis? *J Mol Cell Cardiol* 50(2): 266–72.

Rios M, Williams DA. (1990) Systematic analysis of the ability of stromal cell lines derived from different murine adult tissues to support maintenance of hematopoietic stem cells in vitro. *J Cell Physiol* 145(3): 434–43.

Rosenblatt-Velin N, Lepore MG, Cartoni C, Beermann F, Pedrazzini T. (2005) FGF-2 controls the differentiation of resident cardiac precursors into functional cardiomyocytes. *J Clin Invest* 115(7): 1724–33.

Rubart M, Field LJ. (2006) Cardiac regeneration: repopulating the heart. *Annu Rev Physiol* 68: 29–49.

Rzucidlo EM, Martin KA, Powell RJ. (2007) Regulation of vascular smooth muscle cell differentiation. *J Vasc Surg* 45 Suppl. A: A25–32.

Schönig K, Schwenk F, Rajewsky K, Bujard H. (2002) Stringent doxycycline dependent control of CRE recombinase in vivo. *Nucleic Acids Res* 30(23): e134.

Shin DM, Zuba-Surma EK, Wu W, *et al.* (2009) Novel epigenetic mechanisms that control pluripotency and quiescence of adult bone marrow-derived Oct4(+) very small embryonic-like stem cells. *Leukemia* 23(11): 2042–51.

Shin DM, Liu R, Klich I, Ratajczak J, Kucia M, Ratajczak MZ. (2010) Molecular characterization of isolated from murine adult tissues very small embryonic/epiblast like stem cells (VSELs). *Mol Cells* 29(6): 533–8.

Siu CW, Liao SY, Liu Y, Lian Q, Tse HF. (2010) Stem cells for myocardial repair. *Thromb Haemost* 104(1): 6–12.

Smadja DM, Gaussem P, Mauge L, *et al.* (2009) Circulating endothelial cells: a new candidate biomarker of irreversible pulmonary hypertension secondary to congenital heart disease. *Circulation* 119(3): 374–81.

Smart N, Risebro CA, Melville AA, *et al.* (2007) Thymosin beta4 induces adult epicardial progenitor mobilization and neovascularization. *Nature* 445(7124): 177–82.

Smart N, Riley PR. (2009) Derivation of epicardium-derived progenitor cells (EPDCs) from adult epicardium. In: *Current Protocols in Stem Cell Biology*, Chapter 2:Unit2C.2. New York: Wiley.

Smart N, Risebro CA, Clark JE, *et al.* (2010) Thymosin beta4 facilitates epicardial neovascularization of the injured adult heart. *Ann NY Acad Sci* 1194: 97–104.

Smart N, Bollini S, Dubé KN, *et al.* (2011) De novo cardiomyocytes from within the activated adult heart after injury. *Nature* 2011, 8 June. doi: 10.1038/nature10188. [Epub ahead of print].

Smedley D, Salimova E, Rosenthal N. (2010) Cre recombinase resources for conditional mouse mutagenesis. *Methods* 2010, 31 December. [Epub ahead of print].

Smith RR, Barile L, Cho HC, *et al.* (2007) Regenerative potential of cardiosphere-derived cells expanded from percutaneous endomyocardial biopsy specimens. *Circulation* 115(7): 896–908.

Snippert HJ, van der Flier LG, Sato T, *et al.* (2010) Intestinal crypt homeostasis results from neutral competition between symmetrically dividing Lgr5 stem cells. *Cell* 143(1): 134–44.

Soonpaa MH, Kim KK, Pajak L, Franklin M, Field LJ. (1996) Cardiomyocyte DNA synthesis and binucleation during murine development. *Am J Physiol* 271(5, Pt 2): H2183–9.

Soonpaa MH, Field LJ. (1998) Survey of studies examining mammalian cardiomyocyte DNA synthesis. *Circ Res* 83(1): 15–26.

Soriano P. (1999) Generalized lacZ expression with the ROSA26 Cre reporter strain. *Nat Genet* 21(1): 70–1.

Speer MY, Yang HY, Brabb T, *et al.* (2009) Smooth muscle cells give rise to osteochondrogenic precursors and chondrocytes in calcifying arteries. *Circ Res* 104(6): 733–41.

Stamm C, Choi YH, Nasseri B, Hetzer R. (2009) A heart full of stem cells: the spectrum of myocardial progenitor cells in the postnatal heart. *Ther Adv Cardiovasc Dis* 3(3): 215–29.

Stary E, Gaupp R, Lechner S, *et al.* (2010) New architectures for Tet-on and Tet-off regulation in *Staphylococcus aureus*. *Appl Environ Microbiol* 76(3): 680–7.

Stintzing S, Ocker M, Hartner A, Amann K, Barbera L, Neureiter D. (2009) Differentiation patterning of vascular smooth muscle cells (VSMC) in atherosclerosis. *Virchows Arch* 455(2): 171–85.

Takahashi K, Yamanaka S. (2006) Induction of pluripotent stem cells from mouse embryonic and adult fibroblast cultures by defined factors. *Cell* 126(4): 663–76.

Tallini YN, Greene KS, Craven M, *et al.* (2009) c-kit expression identifies cardiovascular precursors in the neonatal heart. *Proc Natl Acad Sci USA* 106(6): 1808–13.

Tang W, Zeve D, Suh JM, *et al.* (2008) White fat progenitor cells reside in the adipose vasculature. *Science* 322(5901): 583–6.

Tateishi K, Ashihara E, Takehara N, *et al.* (2007) Clonally amplified cardiac stem cells are regulated by Sca-1 signaling for efficient cardiovascular regeneration. *J Cell Sci* 120(10): 1791–800.

Tateishi K, Ashihara E, Honsho S, *et al.* (2007) Human cardiac stem cells exhibit mesenchymal features and are maintained through Akt/GSK-3beta signaling. *Biochem Biophys Res Commun* 352(3): 635–41.

Tateishi K, Takehara N, Matsubara H, Oh H. (2008) Stemming heart failure with cardiac- or reprogrammed-stem cells. *J Cell Mol Med* 12(6A): 2217–32.

Tillmanns J, Rota M, Hosoda T, *et al.* (2008) Formation of large coronary arteries by cardiac progenitor cells. *Proc Natl Acad Sci USA* 105(5): 1668–73.

Tong C, Li P, Wu NL, Yan Y, Ying QL. (2010) Production of p53 gene knockout rats by homologous recombination in embryonic stem cells. *Nature* 467(7312): 211–3.

Torsney E, Xu Q. (2011) Resident vascular progenitor cells. *J Mol Cell Cardiol* 2011 February; 50(2): 304–11. Epub 2010, 17 September.

Uchida S, Gellert P, Braun T. (2011) Deeply dissecting stemness: making sense to non-coding RNAs in stem cells. *Stem Cell Rev* 2011 Jun 25. [Epub ahead of print]. DOI: 10.1007/s12015-011-9294-y.

Urbanek K, Quaini F, Tasca G, *et al.* (2003) Intense myocyte formation from cardiac stem cells in human cardiac hypertrophy. *Proc Natl Acad Sci USA* 100(18): 10440–5.

Urlinger S, Baron U, Thellmann M, Hasan MT, Bujard H, Hillen W. (2000) Exploring the sequence space for tetracycline-dependent transcriptional activators: novel mutations yield expanded range and sensitivity. *Proc Natl Acad Sci USA* 97(14): 7963–8.

Van de Lest CH, Versteeg EM, Veerkamp JH, Van Kuppevelt TH. (1995) Elimination of autofluorescence in immunofluorescence microscopy with digital image processing. *J Histochem Cytochem* 43(7): 727–30.

Verzi MP, McCulley DJ, De Val S, Dodou E, Black BL. (2005) The right ventricle, outflow tract, and ventricular septum comprise a restricted expression domain within the secondary/anterior heart field. *Dev Biol* 287(1): 134–45.

Waldo KL, Kumiski DH, Wallis KT, *et al.* (2001) Conotruncal myocardium arises from a secondary heart field. *Development* 128(16): 3179–88.

Wang X, Hu Q, Nakamura Y, *et al.* (2006) The role of the sca-1+/CD31– cardiac progenitor cell population in postinfarction left ventricular remodeling. *Stem Cells* 24(7): 1779–88.

Watt AJ, Battle MA, Li J, Duncan SA. (2004) GATA4 is essential for formation of the proepicardium and regulates cardiogenesis. *Proc Natl Acad Sci USA* 101(34): 12573–8.

Wen Z, Zheng S, Zhou C, Wang J, Wang T. (2010) Repair mechanisms of bone marrow mesenchymal stem cells in myocardial infarction. *J Cell Mol Med* 2010, 28 December. doi: 10.1111/j.1582–4934.2010.01255.x. [Epub ahead of print].

Wessels A, Pérez-Pomares JM. (2004) The epicardium and epicardially derived cells (EPDCs) as cardiac stem cells. *Anat Rec A Discov Mol Cell Evol Biol* 276(1): 43–57.

Wohlschlaeger J, Levkau B, Brockhoff G, *et al.* (2010) Hemodynamic support by left ventricular assist devices reduces cardiomyocyte DNA content in the failing human heart. *Circulation* 121(8): 989–96.

Wu SM, Fujiwara Y, Cibulsky SM, *et al.* (2006) Developmental origin of a bipotential myocardial and smooth muscle cell precursor in the mammalian heart. *Cell* 127(6): 1137–50.

Wurmser AE, Gage FH. (2002) Stem cells: cell fusion causes confusion. *Nature* 416(6880): 485–7.

Zaruba MM, Soonpaa M, Reuter S, Field LJ. (2010) Cardiomyogenic potential of C-kit(+)-expressing cells derived from neonatal and adult mouse hearts. *Circulation* 121(18): 1992–2000.

Zeisberg EM, Tarnavski O, Zeisberg M, *et al.* (2007) Endothelial-to-mesenchymal transition contributes to cardiac fibrosis. *Nat Med* 13(8): 952–61.

Ziegelhoeffer T, Fernandez B, Kostin S, *et al.* (2004) Bone marrow-derived cells do not incorporate into the adult growing vasculature. *Circ Res* 94(2): 230–8.

7

Conclusion

First of all, I would like to thank you for purchasing and reading my book. I hope I have convinced you that the *in silico* screening method described here is an effective way to predict the laws of nature. The programming involved need not be something that only computer scientists and programmers can do; by using a script language such as Perl, it is easy to perform a task (e.g. finding a set of genes expressed highly in heart across species), which would take ages by hand. In Chapter 2, I surveyed useful databases and datasets that you can use to test your hypothesis. As data are being generated everyday by researchers around the world, there will be more databases and datasets that I have not cited. The best place to keep your eyes on is PubMed. The My NCBI function allows this to be done easily.

As I have repeatedly stressed, there are so many biological (both experimental and computational) data available through the internet. How much money has been spent generating such data and why are they not being used more often? This was the motivation behind me, together with my students and collaborators, initiating a project called the '1,000 Genes Project' to facilitate annotating uncharacterized genes. I have organized my group so that two students are working on the *in silico* screening using Bioinformatics (Mr Pascal Gellert & Mr David John) (see Chapters 3 and 4); Ms Katharina Jenniches is employing antisense morpholino-based knockdown in zebrafish embryos to generate evidence for the necessity for the predicted genes *in vivo* (Chapter 5); Ms Mizue Teranishi is working with murine embryonic stem cells to perform functional assays *in vitro* (Chapter 5); and Ms Piera De Gaspari is utilizing a novel lineage tracing murine system for adult stem cells to establish an assay to test the significance of the identified evolutionarily conserved, tissue-enriched, uncharacterized genes to propose a way to induce adult stem cells to regenerate the damaged body parts (Chapter 6). As there are only 24 hours in a day and we are a group of just six, we cannot do everything by ourselves. To this end, I wrote this book to introduce readers to our approach to further extend our initial efforts to annotate uncharacterized genes. I have tried to be as detailed as possible for each step of our 1,000 Genes Project so that even newcomers can make predictions and validate them experimentally.

I like to think about a scientific project as an investment. An investment can be for scientific advancement, understanding what is not known and most importantly for your future. Dr Abraham Loeb, Chair of the Astronomy department and Director of the Institute for Theory and Computation (ITC) at Harvard University, wrote an interesting article entitled 'The right kind of risk' in *Nature* recently (Loeb, 2010). In it, he describes his view about research-time investment of young astrophysicists using an analogy with economic investments. He uses the following terms: 'bonds' (low risk), 'stocks' (medium risk) and 'venture capital (VC)' (high risk). He recommends that a young researcher (post-doc) diversifies his or her academic portfolio to 50% bonds, 30% stocks and 20% VC. In fact, such an allocation of risks can be performed easily using *in silico* screening. Annotating of completely functionally unknown genes or ncRNAs would equate to VC, whereas feeding in additional components to the existing signaling pathways by means of annotating novel genes can be regarded as bonds and stocks, depending on how well conserved and characterized is the pathway. Of course, this is all a matter of taste.

Working with computers (i.e. writing programs) requires much less time than performing biological experiments, at least to obtain tentative results. At the 19th Annual International Conference on Intelligent Systems for Molecular Biology (ISMB) and 10th European Conference on Computational Biology (ECCB), Dr Luis Serrano described this point very clearly: 'It is hard to recruit a post-doc who is willing to spend the next three years performing biochemical assays for one protein to be published in a mediocre journal but is really necessary one to be done, while another post-doc at the next bench is running high-throughput, functional genomics experiments, which will be published in a high impact factor journal within the next three years.' I completely agree. Although both studies are important for advancing the science, 'hard-core' Biology is often avoided in the early scientific career of a young student or post-doc. However, such biological experiments and findings are fundamental for further advancements of computational predictions. By using our *in silico* screening approach, I hope that we can provide a convenient way to narrow down a list of interesting genes to be studied further.

Whenever I describe our approach, some people tell me that the whole scheme resembles science fiction. I reply: 'Yes, you are right. However, our screening is based on the datasets generated by biologists through biological experiments. It is not that we have simulated some numbers or used mathematical formulas to generate hypothetical datasets. Simply put, our *in silico* screening is a way to extract knowledge from information, i.e. the biological data. If it is only prediction that we have done, then the whole approach is paper science. However, we have validated our screening by biological experiments. To add knowledge to the scientific community, we focused mainly on functionally unknown genes. If you are still in doubt, try our approach yourself. If you are put off by computer programming, simply access our knowledge database "C-It" by selecting your tissue of interest. With one mouse click, our database will return you a list of evolutionarily conserved, tissue-enriched, uncharacterized genes. You can also access other databases to see if you would like to perform your own experiments. At least to me, assigning functions to a novel gene sounds like a good project for a student.' I hope you agree.

Reference

Loeb A. (2010) The right kind of risk. *Nature* 467: 358.

Index

FSC
www.fsc.org
MIX
Papier | Fördert gute Waldnutzung
FSC® C083411

Kolibri 360 GmbH
Ferdinand-Jühlke-Straße 7
99095 Erfurt, Deutschland
produktsicherheit@kolibri360.de